SpringerBriefs in Physics

SpringerBriefs in Physics are a series of slim high-quality publications encompassing the entire spectrum of physics. Manuscripts for SpringerBriefs in Physics will be evaluated by Springer and by members of the Editorial Board. Proposals and other communication should be sent to your Publishing Editors at Springer.

Featuring compact volumes of 50 to 125 pages (approximately 20,000–45,000 words), Briefs are shorter than a conventional book but longer than a journal article. Thus, Briefs serve as timely, concise tools for students, researchers, and professionals.

Typical texts for publication might include:

- A snapshot review of the current state of a hot or emerging field
- A concise introduction to core concepts that students must understand in order to make independent contributions
- An extended research report giving more details and discussion than is possible in a conventional journal article
- A manual describing underlying principles and best practices for an experimental technique
- An essay exploring new ideas within physics, related philosophical issues, or broader topics such as science and society

Briefs allow authors to present their ideas and readers to absorb them with minimal time investment.

Briefs will be published as part of Springer's eBook collection, with millions of users worldwide. In addition, they will be available, just like other books, for individual print and electronic purchase.

Briefs are characterized by fast, global electronic dissemination, straightforward publishing agreements, easy-to-use manuscript preparation and formatting guidelines, and expedited production schedules. We aim for publication 8–12 weeks after acceptance.

More information about this series at http://www.springer.com/series/8902

Ulf W. Gedde

Essential Classical Thermodynamics

 Springer

Ulf W. Gedde
Fibre and Polymer Technology
School of Engineering Sciences in Chemistry
Biotechnology and Health
KTH Royal Institute of Technology
Stockholm, Sweden

ISSN 2191-5423   ISSN 2191-5431   (electronic)
SpringerBriefs in Physics
ISBN 978-3-030-38284-1   ISBN 978-3-030-38285-8   (eBook)
https://doi.org/10.1007/978-3-030-38285-8

This Springer imprint is published by the registered company Springer Nature Switzerland AG
The registered company address is: Gewerbestrasse 11, 6330 Cham, Switzerland

Preface

Even before writing our recent book *Fundamental Polymer Science* (Springer Nature (2019)), I realized that a thorough understanding of thermodynamics is essential if the reader is to be able to capture the content of this specialized field. At this stage, the plan was to prepare a brief account as a supplementary chapter. When the polymer science book was delivered, however, in order to reach the broadest possible audience for a concise and highly readable primer on thermodynamics, David Packer of Springer Nature proposed that I should write a separate book to be included in a series of publications called Springer Brief. At this stage, I also realized that many other books in chemistry, physics, materials science and engineering had the same "problem"; a knowledge about basic classical thermodynamics is required in order to be able to absorb the material in these texts. There are many comprehensive books on thermodynamics. Most of them are 500 pages or longer, and they also have a broader perspective, including kinetics, basic physics, applied thermodynamics, etc. Concise texts on classical thermodynamics are few, and many of them are quite old. These short texts are very useful for newcomers, typically first or second year university students, as a complement to the more comprehensive books. Ph.D. students, postdocs and engineers in industry and research institutes have had their basic thermodynamics training 5–10 years back, and it is possible that some of the thermodynamics is forgotten. This book is essentially aimed at this group. For the last decade, I have been teaching the thermodynamics course at KTH Royal Institute of Technology in Stockholm for students in chemistry and biosciences. This 10-year effort has taught me the problems with understanding thermodynamics. Our graduation has increased by a factor of 3 during this period simply by addressing these issues and providing second- and third-level opportunities during the course. The lesson is that inspiration, concern and time make most students knowledgeable. I feel relaxed and happy now in delivering this text to hopefully many readers. Feel free to mail me if you have questions and ideas concerning the book.

This book has very few references. It focuses to provide only the essentials for the understanding of the subject. I have included quite a few equations and derivations of important expressions (laws). The step to apply the formulae to your problems is not so far-reaching.

Professors Mikael Hedenqvist, Eric Tyrode and Fritjof Nilsson, all at KTH Royal Institute of Technology, Stockholm, Sweden, have provided feedback on the text. Dr. Anthony Bristow has corrected the English. Bristow is a person with a unique combination of sense of language and factual knowledge of science. David Packer of Springer Nature has provided inspiration and great ideas concerning this book and the recent polymer science books. Thank you, David.

Stockholm, Sweden   Ulf W. Gedde
October 7, 2019

Contents

Chapter 1
An Introduction to Thermodynamics and the First Law

Classical thermodynamics is based on several robustly formulated statements, referred to as the laws of thermodynamics. The words of Albert Einstein say something important about thermodynamics: "A theory is the more impressive the greater the simplicity of its premises, the more varied the kinds of things that it relates and the more extended the area of its applicability. Therefore, classical thermodynamics has made a deep impression on me. Thermodynamics is the only science about which I am fairly convinced that, within the framework of the applicability of its basic principles, it will never be overthrown". Another respected scientist, Arnold Sommerfeld, wrote: "Thermodynamics is a funny subject. The first time you go through the subject, you do not understand it at all. The second time, you think you understand it, except for one or two small points. The third time, you know that you do not understand it, but you are so used to the subject that it does not bother you anymore".

The *history* of the establishment of the *first law of thermodynamics* involved persons studying and thinking about energy over a 60-year period. Several important experiments were carried out at the end of the eighteenth century and the first half of the nineteenth century, the most prominent being made by the British American-born Benjamin Thompson, the British Humphry Davy, the German natural scientist Julius Robert Mayer and the brilliant experimentalist James Prescott Joule (England). Mayer, working at the time as a doctor on a ship sailing around the world, noticed the unusual colour of venous blood of shipmen when staying in the warm equatorial parts of the world. This observation started his obsession in thinking about energy, work and heat. It led ultimately to a scientific paper published in 1842, in which the first law of thermodynamics was formulated as an idea. It was Joule who produced the precise experimental evidence, which really established the first law. Joule published his findings in 1843 (first in a preliminary form), in 1847 (the paper was rejected after submission) and finally in 1849. This important paper had the title "On the mechanical equivalent of heat". The first law expressed in mathematical terms was delivered in slightly different forms by several scientists,

U. W. Gedde, *Essential Classical Thermodynamics*, SpringerBriefs in Physics,
https://doi.org/10.1007/978-3-030-38285-8_1

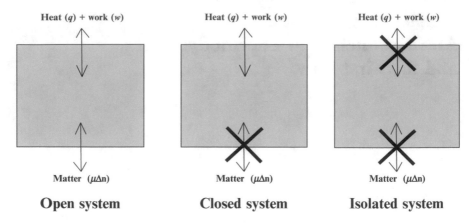

Fig. 1.1 Schematic representation of three different thermodynamic systems

Hermann von Helmholtz (1847), Rudolf Clausius (1850) and William Rankine (1850).

The *first law* states that energy cannot be produced; it can only be transformed from one form to another or, differently formulated, transferred from or to the surroundings of a given system (Fig. 1.1). The energy of the universe, which is an isolated system, is constant. Outside the universe nothing exists, and there can therefore be no transfer of energy to and from the universe.

Figure 1.1 shows three different types of system:

- An *open system* (like the body of a living human) can exchange heat (q), work (w) and matter ($\mu \Delta n$) with its surroundings.
- A *closed system* cannot exchange matter but certainly heat and work with the surroundings. An example of a closed system is a tight metal container.
- An *isolated system* cannot exchange energy in any of the three transfer forms. As already noted, an example is the universe. A perfect thermos flask is another example.

The following equations are mathematical formulations of the first law. Equation (1.1) describes an open system, Eq. (1.2) a closed system and Eq. (1.3) the constant internal energy of an isolated system, e.g. the universe:

$$dU = dq + dw + \mu dn; \Delta U = q + w + \mu \Delta n \qquad (1.1)$$

$$dU = dq + dw; \Delta U = q + w \qquad (1.2)$$

$$dU = 0; \Delta U = 0 \qquad (1.3)$$

where dU is the total differential of the *internal energy* (unit: joule, J), dq and dw are the differential quantities of respectively *heat* and *work* (unit of both: joule, J), μ is the *chemical potential* (unit: joule per mole, J mol^{-1}) and dn is the differential change in the number of moles of matter. An alternative way of describing these

Expansion (change in volume)

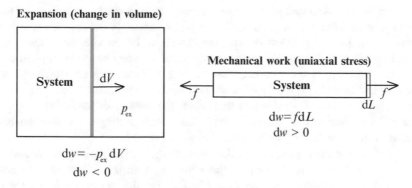

Fig. 1.2 Work performed on or by a system: pressure-volume work with motion of a piston (left) and uniaxial mechanical work (right)

changes (Eqs. (1.1, 1.2, and 1.3)) is $\Delta U = U_2 - U_1$, i.e. the difference in internal energy between states denoted 1 and 2. This difference can also be expressed as $\Delta U = \int dU$ (integration between states 1 and 2). The other parameters are defined according to $q = \int dq$, $w = \int dw$ and $\mu \Delta n = \int \mu dn$, and the integration is performed between states 1 and 2.

The internal energy includes both potential (U_{pot}) and kinetic energy (U_{kin}) (Eq. (1.4)). Please note that the coordinate system is located within the actual system. The internal energy of a container of gas at a certain combination of pressure (p, unit: pascal, Pa), volume (V, unit: cubic metre, m^3) and temperature (T, unit: kelvin, K) is the same on an aeroplane as on the ground. The velocity of the aeroplane should not be added to all the gas molecules in the container. In both cases, the coordinate systems are internal. The kinetic energies include the energies involved in translational motion, rotation and internal motion such as torsions about sigma bonds and vibrations. The potential energy originates from the force fields associated with the primary and secondary bonds. This is how chemistry, i.e. the establishment of new chemical bonds that cause a decrease in the potential energy, has an impact on the temperature in an isolated system (cf. Eqs. (1.3) and (1.4)):

$$U = U_{pot} + U_{kin} \qquad (1.4)$$

Classical thermodynamics defines an addition of the three transferring forms (*heat, work* and *mass transport*) as positive when it adds energy to the system. Heat is positive ($q > 0$) when it is transferred from an external high temperature source. Compression of the matter in a system, $dV < 0$, means that work is done on the system, i.e. $w > 0$. A simple way of remembering the sign is to consider that if you need to do muscle work on the system in order to carry out the change, then $w > 0$.

Figure 1.2 (left) shows gas expansion. In this case, $dV > 0$ and $dw < 0$, i.e. the system is performing work on the surroundings. The differential work (small increment of added work), dw, is the product of force (f) and displacement (dl):

$dw = fdl$. Work associated with a change in volume of a gas is caused by the force that originates from the pressure (p): $f = pA$, where A is the area on which the pressure acts. In Fig. 1.2 (left sketch), the expansion of the volume of the system at the expense of the external volume requires that a force f has to be applied equal to $p_{ex}A$, where p_{ex} is the external pressure and A is the area on which p_{ex} acts. The differential added quantity of work is thus $dw = -p_{ex}Adl$, where $Adl = dV$. The minus sign is due to the fact that when the system expands ($dV > 0$), the system is performing work on the surroundings. Unit analysis: pressure·volume $= Pa \cdot m^3 = N \cdot m^{-2} \cdot m^3 = N \cdot m = J$.

Assuming that the gas expansion sketched in Fig. 1.2 (left) is isothermal and that the gas is ideal, obeying the general gas equation, then the following differential work is done, assuming a *reversible* expansion, i.e. the expansion is carried out extremely slowly in order to maintain mechanical equilibrium throughout the process, viz. $p = p_{ex}$, where p is the system pressure and p_{ex} is the pressure outside the system (external pressure):

$$dw = -p_{ex}dV$$
$$p = p_{ex} \text{ (reversible)} \Rightarrow dw = -pdV \tag{1.5}$$

The first relationship in Eq. (1.5) is general and valid for both irreversible and reversible processes. The second expression is valid only for a *reversible* process. The pressure depends on the volume of the system, according to the general gas equation for an ideal gas. Therefore, the following expression holds for *isothermal* conditions:

$$p = \frac{nRT}{V} \Rightarrow dw = -nRT \cdot \frac{dV}{V} \Rightarrow w = nRT \ln\left(\frac{V_1}{V_2}\right) \tag{1.6}$$

Hence, the work is positive if the final volume (V_2) is smaller than the initial volume (V_1), i.e. for compression, whereas the work is negative for gas expansion ($V_2 > V_1$). The differential work carried out on a stretched band of a *Hookean solid* is given by (Fig. 1.2, right-hand sketch):

$$dw = fdL \tag{1.7}$$

Hooke's law ($\sigma = E\varepsilon$) and the definitions of stress ($\sigma = f/A_0$, where f is the force, A_0 is the cross-sectional area) and strain (ε), which is equal to $(L - L_0)/L_0$, where L is the length of the stressed specimen and L_0 is the length of the unstressed specimen, yield the following expressions:

Fig. 1.3 Stress-strain curve of a typical metallic material with an elastic, Hookean performance at low strains. The grey area is the work done on the Hookean material per unit volume of matter

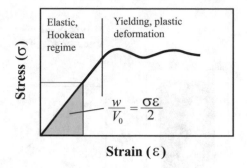

Strain (ε)

$$\sigma = E\varepsilon \Rightarrow f = \frac{EA_0}{L_0} \cdot (L - L_0)$$

$$w = \frac{EA_0}{L_0} \cdot \int_{L_0}^{L} (L - L_0)dL = \left| \frac{EA_0}{L_0} \left(\frac{L^2}{2} - L_0 L \right) \right|_{L_0}^{L} = \tag{1.8}$$

$$\frac{1}{2} \cdot \frac{EA_0}{L_0} \cdot (L - L_0)^2 = \frac{1}{2} \cdot \frac{L - L_0}{L_0} \cdot \frac{L - L_0}{L_0} \cdot E \cdot A_0 \cdot L_0$$

$$w = \frac{\sigma \varepsilon}{2} \cdot V_0$$

Equation (1.8) shows that the work per unit volume is the integral of the stress-strain curve (Fig. 1.3).

Most *calorimetric studies* are carried out at constant pressure. Our daily activities are performed at almost *constant pressure* (1 atm $\approx$ 0.101 MPa), and this is the reason for the introduction of a help state function, the *enthalpy* (H), which is defined as:

$$H = U + pV \tag{1.9}$$

Enthalpy has the unit joule, J. By differentiating H and combining this expression with Eqs. (1.2) and (1.5), it is shown that the *total differential* of the *enthalpy* is equal to the *different heat flow* at constant pressure:

$$dH = dU + pdV + Vdp \wedge dU = dq - pdV$$
$$dH = dq - pdV + pdV + Vdp \tag{1.10}$$
$$dH = dq \ (p = \text{constant})$$

The difference in enthalpy between two states, both at the same pressure, $\Delta H = H_2 - H_1$, is equal to the total heat flow (q). The latter can be accurately measured in a device referred to as a *calorimeter*. The first calorimeters, which used large samples, were made in the last half of the eighteenth century by the chemists

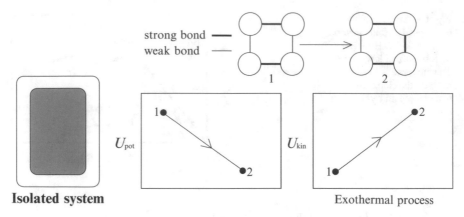

Fig. 1.4 Schematic representation of a chemical reaction strongly binding two molecules within an isolated system. The effects on the potential and kinetic energies are sketched

Joseph Black and Antoine Lavoisier. Modern calorimeters with very high precision are called *differential scanning calorimeters* (abbreviated *DSC*). These instruments provide precise data using a small amount of sample (typically 1–20 mg). This is why enthalpy is such an important quantity.

Internal energy (U) has two "faces", which can be explained by using atoms and molecules as vehicles, the *potential energy* (U_{pot}) due to the force fields between atoms and the *kinetic energy* (U_{kin}) due to the motion of molecules. Two instructive examples are presented. Figure 1.4 shows a *chemical reaction* between two diatomic molecules. A strong bond is formed between the two molecules, replacing a weak bond. The effect is that the potential energy of the system decreases. Why? The potential energy is zero when two atoms are far from each other and are unable to sense the presence of the other atom. As the atoms come closer, the potential energy decreases until a minimum is reached. A strong bond means that the potential energy of the two-atom system is very low (a highly negative value), and this is why the replacement of a weak bond by a strong bond causes a decrease in the potential energy. The system under consideration is isolated, i.e. U is constant. Hence, if U_{pot} decreases, U_{kin} must increase by the same quantity (cf. Eq. (1.4)). The increase in kinetic energy means that the temperature of the system increases, $\Delta T > 0$, and the process is referred to as *exothermal*. Let us consider the isothermal case, with the same reaction. The system boundary is changed from being insulating to being thermally conductive, and in order to keep the temperature constant, the increased kinetic energy has to be transported from the system to the surroundings, viz. $\Delta U < 0$. At constant pressure, the enthalpy is used to describe the changes in the system. An exothermal process is characterized by a negative value of ΔH. Assume that the reaction shown in Fig. 1.4 is going backwards, i.e. a weak bond is replacing a strong bond. If the process is *adiabatic* (meaning "without heat transfer", i.e. q is strictly zero), the temperature decreases, $\Delta T < 0$, and under isothermal conditions, both ΔU (constant V) and ΔH (constant p) take positive values. This type of process is referred to as *endothermal*. A process that is characterized by both ΔU (constant V) and ΔH (constant p) being equal to zero is referred to as *athermal*.

Another example is the *elastic deformation of metals and rubbers*. Both these materials are elastic; i.e. none of the deformation energy is permanently transformed into heat. We drop a metal ball and a rubber ball from a height of 1 m. The balls fall and hit the ground. In their contact with the ground, the balls deform (flatten), and after a very short period of time, they leave the ground and bounce upwards. Suppose that we leave the room for a cup of coffee, say for 15 min. Suppose further that we removed all the air in the container where the balls are bouncing. We sense after our return not only pleasure with the refreshing effect of the coffee but also with the observation that the balls are still bouncing. Both metal and rubber behave in the same way, but with one important difference. When the rubber ball is in contact with the ground, the temperature of the rubber increases. Immediately, after leaving the ground, the temperature is the same as before the bouncing. The metal ball maintains a constant temperature throughout the cycle of the bouncing experiment. The work done on the rubber causes a temperature increase, U_{kin} increases and U_{pot} is unchanged, whereas the work done on the metal results in an increase in U but only in U_{pot}. Rubbers show a different elasticity than metals; they are *entropy-elastic*. A simple experiment that you can carry out yourself is that done by the British natural philosopher John Gough in 1805. Stretch a small piece of rubber three to five times its original length. Move the stretched rubber to your lips and you will sense a temperature increase. If you allow the rubber to retract to its original length, the warm feeling from the rubber on your lips disappears. The work done by your muscles on the rubber band is converted to an increase in temperature of the rubber.

Let us introduce a few formal aspects, first focusing on the internal energy (U). The internal energy is a *state function*, with a value depending strictly only on the actual values of the quantities which characterize the state, such as the temperature (T), volume (V), pressure (p) and moles of particles (n). Let us select a closed system, so that n is constant. The ideal gas law, $p = nRT/V$, means that p is dependent only on T and V. Hence, U is a function only of the independent parameters T and V, and the following differential applied:

$$dU = \left(\frac{\partial U}{\partial T}\right)_V dT + \left(\frac{\partial U}{\partial V}\right)_T dV = C_V dT + \pi_T dV \qquad (1.11)$$

where C_v is the *heat capacity at constant volume* (unit: joule per kelvin, J K^{-1}) and π_T is the *internal pressure* (unit: pascal, Pa). For an ideal gas, π_T is zero. For other phases, it is normally large and positive at moderate pressure. The enthalpy (also a state function) is often described in a similar fashion based on $H = f(T, p)$:

$$dH = \left(\frac{\partial H}{\partial T}\right)_p dT + \left(\frac{\partial H}{\partial p}\right)_T dp = C_p dT + \left(\frac{\partial H}{\partial p}\right)_T dp \qquad (1.12)$$

Fig. 1.5 Schematic representation of the enthalpy of reactants and products. Note the two different paths (denoted A and B) to reach point "stop" from point "start"

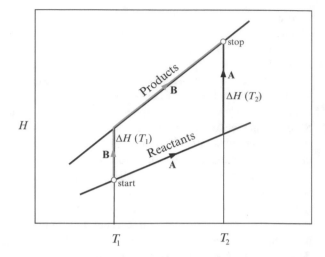

where C_p is the *heat capacity at constant pressure* (unit: joule per kelvin, $J\,K^{-1}$). The difference $C_p - C_v$ is large for an ideal gas, nR; hence, the difference per mole ($C_{p,m} - C_{v,m}$) is R (unit: joule per kelvin and mole, $J\,K^{-1}\,mol^{-1}$). This relation is sometimes called *Mayer's relation* after Julius Robert Mayer. The difference $C_{p,m} - C_{v,m}$ is small for liquids and solids. The molar heat capacity at constant volume of a monatomic gas (e.g. helium or argon) is $3R/2$, and it has three *degree of freedom* (translation motion along x, y and z). A diatomic gas, like O_2 and N_2, has a further two degrees of freedom due to the rotation, i.e. in all five degrees of freedom and thus $C_{v,m} = 5R/2$. Molar heat capacity data covering wide temperature ranges are often expressed in the following *empirical formula*: $C_{p,m}$ (unit: $J\,K^{-1}\,mol^{-1}$) $= a + bT + cT^{-2}$, where the coefficients a, b and c are unique to the particular compound. The heat capacity at low temperatures (0–20 K) is described by the Debye equation (cf. Chap. 2).

An important relationship is the *Kirchhoff law* (sometimes referred to as the Kirchhoff law of thermochemistry. Gustav Kirchhoff, a German physicist, contributed in several fields and the laws in electronics are famous) by which the temperature dependence of phase transitions and chemical reactions can be calculated based on the heat capacity data of the reactants and products. Figure 1.5 shows how the enthalpy of a reaction, in this case reactant → products, changes as a function of temperature. The slopes of the product and reactant lines are the *heat capacities at constant pressure* (C_p) of these compounds. In practice these lines are slightly curved. The derivation of the Kirchhoff law begins from the fact that H is a state function and that two different paths should lead to the same difference in H. Path A involves heating the reactants to temperature T_2 followed by a transition at T_2, whereas path B starts with a transition at T_1 after which the products are heated to T_2. The following expressions are obtained:

$$\Delta H(T_1) + \int_{T_1}^{T_2} C_{p,p} dT = \Delta H(T_2) + \int_{T_1}^{T_2} C_{p,r} dT$$

$$\Delta H(T_2) = \Delta H(T_1) + \int_{T_1}^{T_2} \left(C_{p,p} - C_{p,r} \right) dT \qquad (1.13)$$

where the indices p and r represent products and reactants, respectively. Equation (1.13) is known as the Kirchhoff law and it is a very useful expression. The enthalpy change associated with a given process is normally reported for only one (standard) temperature. By application of the Kirchhoff law, it can be calculated for any other temperature, provided the heat capacity data are known.

Important obstacles that hinder us from understanding thermodynamics are the concepts of *reversibility* and *irreversibility*. A reversible process is carried out just in our mind by making extremely slow changes in the state functions: T, p, etc. The system is kept at *thermal*, *mechanical* and *chemical equilibrium* throughout the process. The process is infinitely slow, which makes it impossible to carry out as a practical experiment, but very useful in the calculation of state functions such as U and H. The conditions at the start (A) and at the end (B) obviously need to be known. The useful journey for calculation between A to B is a *reversible path*. The word itself means that the path can go either forwards or backwards and in both cases follow a single path. This is only possible if the system is in mechanical, thermal and chemical equilibrium with its surroundings. This idealized and extremely useful case is different from all *real processes*, which are always *irreversible*.

One illustrative case is the *isothermal expansion* of an *ideal gas* from volume V to $2V$. The reversible process of maintaining mechanical equilibrium across the process requires that p(system) $= p$(external) and the ideal gas law is obeyed: $pV = nRT$, which means that the work $w = -\int nRTdV/V$ (limits V and $2V$) and $w = -nRT\ln2$. An ideal gas is characterized by the fact that the gas molecules do not sense each other; $(\partial U/\partial V)_T = 0$ and thus U is a function solely of T; ΔU for any isothermal process of an ideal gas has to be zero, viz. $\Delta U = 0$ and $q = -w = nRT \ln2$. As we shall see in Chap. 2, this increases the *entropy* (S) of the ideal gas by $\Delta S = q/T = nR \ln 2$. The entropy is also a state function, and it is calculated by assuming a reversible path from A (volume $= V$) to B (volume $= 2V$) at constant temperature.

The foregoing paragraph showed that an isothermal reversible change in volume of an ideal gas causes either a positive heat release ($q > 0$) if the gas expands or a negative heat release ($q < 0$) if the gas is compressed. Another historically important case is the *reversible adiabatic volume change* of an ideal gas. Let us consider the *reversible adiabatic compression of an ideal gas*. Compression means that work is done on the system, i.e. $w > 0$. Adiabatic means that $q = 0$. This implies that the volume ($V_1 \rightarrow V_2$) is decreasing and that the temperature ($T_1 \rightarrow T_2$) is increasing. This case is an important part of the *Carnot cycle*, a *foundation of the second law* (cf. Chap. 2). The fact that the process is adiabatic ($dq = 0$) and reversible means that:

Fig. 1.6 Reversible
adiabatic compression of an
ideal gas divided into two
reversible processes.
Process 2 accounts for the
whole change in U

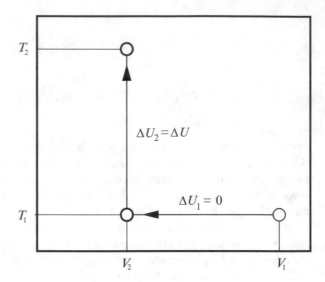

$$dU = dq + dw = dw \wedge dw = -pdV \Rightarrow dU = -pdV \qquad (1.14)$$

The internal energy of an ideal gas is determined only by its temperature according to:

$$dU = C_V dT \qquad (1.15)$$

where C_V is heat capacity at constant volume. The following equation is obtained by combining Eqs. (1.14) and (1.15):

$$C_V dT = -pdV; C_V dT = -nRT \frac{dV}{V} \Rightarrow \left(\frac{C_V}{nR}\right) \cdot \frac{dT}{T} = -\frac{dV}{V} \qquad (1.16)$$

which can be integrated according to:

$$\left(\frac{C_V}{nR}\right) \cdot \int_{T_1}^{T_2} \frac{dT}{T} = -\int_{V_1}^{V_2} \frac{dV}{V} \Rightarrow \ln\left(\frac{T_2}{T_1}\right) = \ln\left(\frac{V_2}{V_1}\right)^{-\frac{nR}{C_V}} \qquad (1.17)$$

$$T_2 = T_1 \cdot \left(\frac{V_2}{V_1}\right)^{-\frac{R}{C_{V,m}}} \qquad (1.18)$$

where $C_{V,m}$ is the molar heat capacity.

The change in U associated with the adiabatic compression of an ideal gas is when C_V is assumed constant, according to Fig. 1.6, given by:

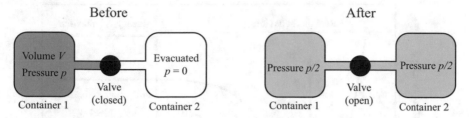

Fig. 1.7 The joule expansion experiment in a schematic form

$$\Delta U = \int_{T_1}^{T_2} C_v dT = C_v T_1 \left(\left(\frac{V_2}{V_1} \right)^{-\frac{R}{C_{v,m}}} - 1 \right) \tag{1.19}$$

Joule expansion (or *free expansion*) is another historical case (Fig. 1.7). The gas is kept in container 1 and container 2 is evacuated. The valve between the two containers is opened and the gas in container 2 expands into container 2. This was an experiment conducted by John Leslie, Joseph Louis Gay-Lussac and the great experimentalist James Prescott Joule in the 1800 century.

Joule expansion is irreversible, because the system is not in balance with its surroundings. The work (w) associated with the expansion is zero, because $p_{ex} = 0$ ($dw = -p_{ex} dV$). The process is rapid and essentially no heat transfer is possible, i.e. the process is approximately adiabatic, $q = 0$. The change in internal energy $\Delta U = q + w = 0$, i.e. the internal energy is constant. The changes in the system as a result of this brutal (irreversible) gas expansion depend on the nature of the gas. The decisive equation is Eq. (1.11), which for convenience is here repeated:

$$dU = \left(\frac{\partial U}{\partial T} \right)_V dT + \left(\frac{\partial U}{\partial V} \right)_T dV - C_v dT + \pi_T dV \tag{1.20}$$

where π_T denotes the internal pressure.

An ideal gas is characterized by the absence of interactions between the gas molecules, i.e. $\pi_T = 0$, which then modifies Eq. (1.20) to:

$$dU = C_v dT \Rightarrow dT = 0 \text{ if } dU = 0 \Rightarrow U = \text{constant} \tag{1.21}$$

which implies that the joule expansion of an ideal gas is isothermal. A real gas has a non-zero internal pressure (Fig. 1.8). At moderately high pressures (attractive region), $\pi_T > 0$, whereas at very high pressures (repulsive region), π_T has negative values. The ideal gas approximation, characterized by $\pi_T = 0$, is valid at low pressures. Under these conditions, the distance between adjacent gas molecules is sufficiently large that their potential energy is zero, and they do not sense the presence of their neighbours.

The internal energy (U) is also constant for a real gas, i.e.:

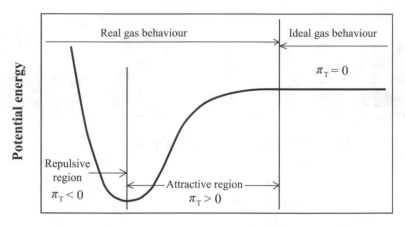

Intermolecular distance

Fig. 1.8 Potential energy as a function of distance between gas molecules (sketch)

$$C_v dT + \pi_T dV = 0 \Rightarrow dT = -\frac{\pi_T}{C_v} \cdot dV \qquad (1.22)$$

When a gas expands ($dV > 0$), noting that C_v is always positive, the accompanying temperature change is negative ($\Delta T < 0$) for a real gas in the attractive region and positive ($\Delta T > 0$) in the repulsive region (cf. Fig. 1.8).

Before discussing the second law, we have to mention the *zeroth law*, which can be expressed as follows: if system A is in thermal equilibrium with system C and system B is in thermal equilibrium with system C, then system A is in thermal equilibrium with system B. The zeroth law is often neglected in textbooks. These relations were expressed in different ways by several scientists, e.g. James Maxwell, Constantin Carathéodory and Max Planck. The name "zeroth law" was coined by Ralph Fowler in the 1930s.

Chapter 2
The Second and Third Laws

The *second law* is probably the most difficult to grasp among the thermodynamic laws. This law states that the *entropy* (i.e. the state function that is emerging from the second law) in the universe is constantly increasing; the reason for the increase being that all real processes are *irreversible*. They do not take place in a gentle reversible fashion with a controlled mechanical, thermal and chemical balance between the system and its surroundings. The fact that heat is transferred from a hot spot to colder surroundings is not described by the first law (cf. Chap. 1). We burn our fingers by touching a hot metallic plate. The second law provides the means to describe this well-known fact. Another reason for the emerging second law was the *steam engine*, with roots in the late seventeenth century (Thomas Savary) and further developed during the eighteenth and early nineteenth centuries (Thomas Newcomen, Mårten Triewald, Jacob Leupold, James Watt, Richard Trevithick, John Smeaton, Oliver Evans, etc.). During the nineteenth century, the steam engine became increasingly important, and the centre of its development was in England and Scotland. The French showed a considerable interest in this industrial development, and they gave Nicolas Léonard *Sadi Carnot*, a young engineer of prominent family, the opportunity to study steam engines. His name Sadi originates from a Persian poet, Sadi of Shiraz. His work resulted in a very important book entitled *Reflections on the Motive Power of Fire*. The year was 1824 and Carnot was only 28 years old. He described a cyclic process sketched in Fig. 2.1, which consists of four reversible processes from which important information was gained and led the two prominent scientists, Rudolf Clausius and William Thomson (Lord Kelvin), to formulate the second law of thermodynamics in the 1850s. Sadi Carnot's life ended when he was only 36 years old. One result of his fatal cholera illness was that almost all of the 600 printed volumes of his book were buried. Luckily a few books were rescued for Clausius, Thomson and others (among them Rudolf Diesel).

The precise mathematical formulation of the second law is presented in the coming text; please, just be patient.

Figure 2.2 shows a simple graphic representation of the *Carnot cycle*, where the energy balance of the engine is described by:

© The Author(s), under exclusive license to Springer Nature Switzerland AG 2020
U. W. Gedde, *Essential Classical Thermodynamics*, SpringerBriefs in Physics,
https://doi.org/10.1007/978-3-030-38285-8_2

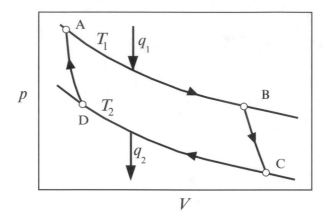

Fig. 2.1 The Carnot cycle with (AB) isothermal expansion (heat ($q_1 > 0$) is transferred from the hot source), (BC) adiabatic expansion, which leads to a decrease in temperature, $T_1 \rightarrow T_2$, (CD) isothermal compression (heat ($q_2 < 0$) is transferred from the engine to the cold source), and (DA) adiabatic compression which causes a temperature increase, $T_2 \rightarrow T_1$). Details about these processes are found in Chap. 1

Fig. 2.2 Simplified sketch of the Carnot cycle. Note that the signs of the different energy transfers to the engine are positive if the arrows are pointing inwards towards the engine

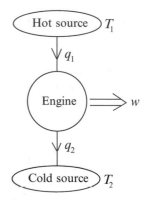

$$q_1 + q_2 + w = 0 \Rightarrow -w = q_1 + q_2 \tag{2.1}$$

where $-w$ is the work performed by the engine. The *efficiency of the engine* (ε) is given by:

$$\varepsilon = \frac{-w}{q_1} = \frac{q_1 + q_2}{q_1} = 1 + \frac{q_2}{q_1} \tag{2.2}$$

The heat flows, q_1 and q_2, are obtained respectively from the isothermal expansion (path AB; Fig. 2.1) and from the isothermal compression (path CD; Fig. 2.1) according to (cf. Chap. 1):

$$q_1 = nRT_1 \ln \left(\frac{V_B}{V_A} \right) \tag{2.3}$$

$$q_2 = nRT_2 \ln \left(\frac{V_D}{V_C} \right) = -nRT_2 \ln \left(\frac{V_B}{V_A} \right) \tag{2.4}$$

By combining Eqs. (2.3) and (2.4), the following important equations are obtained:

$$\frac{q_2}{q_1} = \frac{-nRT_2 \ln \left(\frac{V_B}{V_A} \right)}{nRT_1 \ln \left(\frac{V_B}{V_A} \right)} = -\frac{T_2}{T_1} \Rightarrow \varepsilon = 1 - \frac{T_2}{T_1} \tag{2.5}$$

$$\frac{q_1}{T_1} + \frac{q_2}{T_2} = 0 \Rightarrow \oint \frac{dq_{rev}}{T} = 0 \tag{2.6}$$

Equation (2.5) shows that not even a *perfect engine* can be 100% efficient since T_2 cannot be zero. A fraction of the heat flow from the hot source (q_1) is always transferred further to the cold source (q_2). *Heat cannot be completely converted into work*. A real engine has a lower efficiency than the fully reversible Carnot engine. Why is that? The added effect of friction results in a greater loss of heat to the cold source than is predicted by Eq. (2.4). Equation (2.6) shows that the quantity q_{rev}/T behaves like a state function, i.e. the cyclic integral of $\partial q_{rev}/T$ is zero, and it was given the name *entropy* with the symbol S and with the unit joule per kelvin, J K^{-1}.

Figure 2.3 shows a complex Carnot process, which still obeys the simple mathematics of Eq. (2.6). By this generalized scheme, *any cyclic process* can be represented by a series of Carnot isotherms and adiabatic curves.

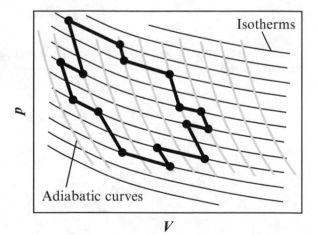

Fig. 2.3 A complex combination of Carnot processes forming a cycle. It consists of eight isothermal processes and eight adiabatic processes

Isotherms

Adiabatic curves

p

V

Fig. 2.4 A cyclic process
with mixed reversible and
irreversible regimes

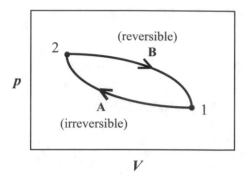

Fig. 2.4 A cyclic process
with mixed reversible and
irreversible regimes

A few more words about practical steam engines: they are often described based on the *Rankine cycle* (named after William John Macquorn Rankine, a Scottish professor of mechanical engineering), and sometimes referred to as the *practical Carnot cycle*. Further information about the Rankine cycle and its applications can be found in textbooks on technical thermodynamics, such as Schmidt (2019). If the arrows of the heat transfer and work in Fig. 2.2 are reversed, heat can be transferred from the cold source to the warm source. The work done on the engine must be supplied by, e.g. electric power. This is the principle of the *heat pump*, which was theoretically described by William Thomson in 1852. The first heat pump was built a few years later by the Austrian engineer Peter von Rittinger.

Figure 2.4 shows a cyclic process which is partly reversible, from 2 to 1 (path B). The process which starts at 1 and ends at 2 is irreversible. The irreversible process has a larger leakage of heat to the cold source, which implies that the $\int dq/T$-contribution is more negative than that of a reversible process. The cyclic integral shown in Eq. (2.7) is therefore negative:

$$\oint \frac{dq}{T} < 0 \tag{2.7}$$

The cyclic integral in Fig. 2.4 is divided into a reversible section (S_1–S_2) and an irreversible section:

$$\int_1^2 \frac{dq}{T} + \int_2^1 \frac{dq_{\mathrm{rev}}}{T} < 0 \tag{2.8}$$

$$\int_1^2 \frac{dq}{T} + S_1 - S_2 < 0 \Rightarrow S_2 - S_1 > \int_1^2 \frac{dq}{T} \therefore dS > \frac{dq}{T} \tag{2.9}$$

Equation (2.9) is an important formulation by Rudolf Clausius, hence referred to as the *Clausius theorem* or the *Clausius inequality principle*, which is valid for any real spontaneous irreversible process. This very compact expression is of *extreme, decisive importance*. It provides the key to what is a *spontaneous process*. It is the basis for the coining of the free energy concepts by Gibbs and Helmholtz.

For a reversible process:

$$dS = \frac{dq_{rev}}{T} \tag{2.10}$$

For an isolated system, e.g. the universe, $dq = 0$ and, according to Eq. (2.9), $dS > 0$. The *entropy of the universe increases with time*. Time cannot be reversed, because that would require entropy to decrease, in violation of the second law.

A new formulation, containing *only state functions* based on both the first and second laws, is thus obtained:

$$dU = dq - pdV + \mu dn = TdS - pdV + \mu dn \tag{2.11}$$

The entropy of some substances is given a precise value by the *third law*:

$$S(T) = S(0) + \Delta S(\text{Debye law}) + \int_{T_d}^{T_m} \frac{C_{p,c}dT}{T} + \frac{\Delta H_m}{T_m} + \int_{T_m}^{T} \frac{C_{p,m}dT}{T} \tag{2.12}$$

where $S(0) = 0$ for a crystalline compound according to the *third law* of thermodynamics (also called the *Nernst heat theorem* after Walther Nernst, German chemist and Nobel laurate in 1920) and $C_{p,c}$ and $C_{p,m}$ are the heat capacities of the crystal phase and the melt, respectively. The third law is less commonly known than the first and second laws. One way of expressing the third law is to say that the entropy of a system approaches a constant value when the temperature approaches zero (0 K). At equilibrium, when a system is approaching 0 K, the energy of the system approaches a minimum value. If only one ground state exists, then the entropy at 0 K is zero: $S(0) = 0$. An example of such system is a 100% crystalline substance with a well-defined low-energy crystal structure. A low plateau value (called the residual entropy) is obtained for systems that lack well-defined order, e.g. a glass or a semicrystalline compound. An alternative expression of the third law was proposed by Nernst: "The entropy change associated with any condensed system undergoing a reversible isothermal process approaches zero as the temperature approaches 0 K". Another statement of Nernst is that "It is impossible for any process, no matter how idealized, to reduce the entropy of a system to its absolute zero value in a finite number of operations".

The *heat capacity law of Debye* (developed in 1912 by Peter Debye, a Dutch-American physical chemist, Nobel laurate in 1936) is used to obtain the integral from 0 to T_d, which can be 20 K:

Fig. 2.5 Crystallization of supercooled water at 253 K, an irreversible process. Division of the irreversible process into a number of reversible subprocesses

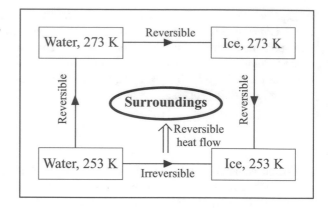

$$C_p = aT^3 \Rightarrow \Delta S(\text{Debye law}) = \int_0^{T_d} \frac{aT^3 dT}{T} = \int_0^{T_d} aT^2 dT \qquad (2.13)$$

where a is a constant.

The second law makes it possible to judge whether or not a process is spontaneous. The *criterion for spontaneity* is based on two quantities: the change in entropy which occurs within the system of consideration as a consequence of the process (ΔS_{syst}) and the effect of the process on the entropy of the surroundings (i.e. the rest of the universe), ΔS_{surr}. The sum of the two changes in entropy, $\Delta S_{syst} + \Delta S_{surr}$, is referred to as the *entropy production* (ΔS_i), and this is positive for a spontaneous process. The following example shows how this principle can be applied. Water at 253 K crystallizes irreversibly. The entropy change associated with crystallization of water is calculated by finding a reversible path (Fig. 2.5): water is heated reversibly from 253 to 273 K. At 273 K, water and ice have the same right to exist thermodynamically (the same Gibbs free energy, a term explained in Chap. 3) and water at 273 K is transformed reversibly into ice. Finally, ice is cooled reversibly from 273 to 253 K. Equation (2.14) presents the sum of the three contributions:

$$\Delta S_{syst} = \int_{253}^{273} \frac{C_{p,l} dT}{T} + \frac{\Delta H_{cryst}}{273} + \int_{273}^{253} \frac{C_{p,c} dT}{T} \qquad (2.14)$$

where $C_{p,l}$ is the heat capacity of liquid water, ΔH_{cryst} is the enthalpy change of crystallization at the equilibrium melting point (273 K) and $C_{p,c}$ is the heat capacity of ice. The actual crystallization process occurs at 253 K and is irreversible. However, the heat is transferred in a reversible fashion to the surroundings with $-\Delta H_{cryst}$ at 253 K, which is equal to the melting enthalpy (ΔH_{melt}) at 253 K:

$$\Delta S_{surr} = \frac{\Delta H_{melt}}{253} \qquad (2.15)$$

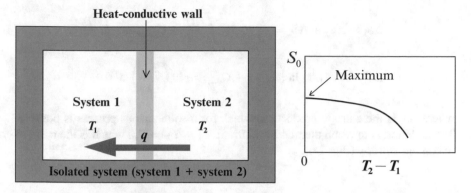

Fig. 2.6 An isolated system consisting of two pieces of matter at different temperatures, $T_2 > T_1$. The two systems are thermally connected by a heat conductive wall. The graph to the left shows the entropy production associated with the process approaching thermal equilibrium ($T_2 = T_1$)

The calculation of ΔH_{melt} at 253 K requires the use of the Kirchhoff law (cf. Chap. 1, Eq. (1.13)). It should be noted that $\Delta S_{\text{syst}} < 0$ ($\Delta H_{\text{cryst}} < 0$) and $\Delta S_{\text{surr}} > 0$. The entropy production (ΔS_i) is positive and is obtained according to:

$$\Delta S_i = \Delta S_{\text{syst}} + \Delta S_{\text{surr}} > 0 \qquad (2.16)$$

Figure 2.6 shows the entropy production (ΔS_i) in the spontaneous process of establishing *thermal equilibrium between two systems* which have initially different temperatures. The entropy production is calculated in two steps. The first task is to calculate the final (equilibrium) temperature (T_{eq}), which requires a knowledge of the initial temperatures, the masses of the two systems and their heat capacities:

$$\begin{aligned}
\Delta H_1 + \Delta H_2 &= 0 \\
C_{\text{p},1,\text{w}} \cdot m_1 \cdot \left(T_{\text{eq}} - T_{1,0}\right) + C_{\text{p},2,\text{w}} \cdot m_2 \cdot \left(T_{\text{eq}} - T_{2,0}\right) &= 0
\end{aligned} \qquad (2.17)$$

where $T_{1,0}$ is the initial temperature of the cold system, $T_{2,0}$ is the initial temperature of the warm system, m_1 and m_2 are respectively the masses of the cold and warm systems and $C_{\text{p},1,\text{w}}$ and $C_{\text{p},2,\text{w}}$ are the heat capacities per unit mass. This calculation is based on the fact that the two systems are contained in a perfect thermos, i.e. together they form an isolated system. It is assumed that the heat capacities of the two are constant, independent of temperature. Both systems have internally uniform temperatures, T_1 and T_2, and the heat flow from warm to cold is through the heat conductive wall. This kind of system would in practice require that each system had a convective means of removing internal temperature gradients.

The second task is to calculate the entropy changes involved in the reversible heating of the cold system to the equilibrium temperature and in the reversible cooling of the warm system to the equilibrium temperature:

$$\Delta S_i = \Delta S_1 + \Delta S_2 = C_{p,1,w} m_1 \int\limits_{T_{1,0}}^{T_{eq}} \frac{dT}{T} + C_{p,2,w} m_2 \int\limits_{T_{2,0}}^{T_{eq}} \frac{dT}{T}$$

$$\Delta S_i = C_{p,1,w} m_1 \ln \left(\frac{T_{eq}}{T_{1,0}} \right) + C_{p,2,w} m_2 \ln \left(\frac{T_{eq}}{T_{2,0}} \right) > 0$$

(2.18)

where ΔS_i is the entropy production, which for a spontaneous process is positive. The ambition is to reach thermal equilibrium, $T_1 = T_2 = T_{eq}$, which is the point of maximum entropy (Fig. 2.6).

Chapter 3
Gibbs and Helmholtz Free Energies

Figure 3.1 illustrates the reason for the founding of two thermodynamic state functions, the *Gibbs free energy* (*G*) and the *Helmholtz free energy* (*A*). The latter is also called the *work function*. Both free energy types, *G* and *A*, have the same unit, joule (J). These useful functions were respectively coined by Josiah Willard Gibbs and Hermann von Helmholtz during the 1870s.

The idea is brilliant in its simplicity. The focus is on the actual process in the actual system. The effect on the surroundings is considered on the basis of the process itself, and the principle described in Chap. 2 regarding the *positive entropy production* is applied. The assumption made, as indicated in Fig. 3.1, is that the temperature and pressure are constant. The change in entropy of the *system* associated with a process is denoted *dS*. The corresponding change in enthalpy of the system is *dH*, which implies that the *surroundings* undergo a change in enthalpy of $-dH$ (the change in enthalpy of the system multiplied by minus one) under isothermal conditions. The entropy change of the surroundings is $-dH/T$. The second law states that the sum of the changes in entropy in both the system and its surroundings should be positive for a spontaneous process (Eq. 2.16):

$$dS - \frac{dH}{T} > 0 \Rightarrow TdS - dH > 0 \tag{3.1}$$

The Gibbs free energy (*G*) is defined according to:

$$G = H - TS \Rightarrow dG = dH - TdS - SdT \tag{3.2}$$

which at constant temperature reduces to:

$$dG = dH - TdS \tag{3.3}$$

By combining Eqs. (3.1) and (3.3), the condition for a spontaneous process is given by:

© The Author(s), under exclusive license to Springer Nature Switzerland AG 2020
U. W. Gedde, *Essential Classical Thermodynamics*, SpringerBriefs in Physics,
https://doi.org/10.1007/978-3-030-38285-8_3

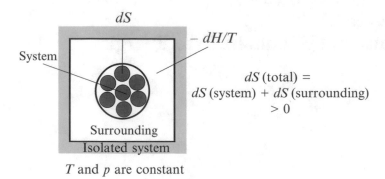

Fig. 3.1 How the second law became the reason for the invention of Gibbs free energy

$$dG < 0 \tag{3.4}$$

At constant T and constant p, the Gibbs free energy (G) strives towards a minimum value. At equilibrium, $dG = 0$, i.e. $G =$ constant. If the temperature and volume are constant, the second law criterion for spontaneity is given by:

$$dS - \frac{dU}{T} > 0 \Rightarrow TdS - dU > 0 \tag{3.5}$$

The Helmholtz free energy (A) is defined according to:

$$A = U - TS \Rightarrow dA = dU - TdS - SdT \tag{3.6}$$

which at constant temperature reduces to:

$$dA = dU - TdS \tag{3.7}$$

and the following criterion for spontaneity at constant T and V is obtained by combining Eqs. (3.5) and (3.7):

$$dA < 0 \tag{3.8}$$

Let us show the usefulness of the free energy concept in an example. Assume that we have an isobaric process, either a chemical reaction or a physical process, A $\rightarrow$ B. The change in Gibbs free energy (ΔG) at a given temperature is then:

$$\Delta G = \Delta H - T\Delta S \tag{3.9}$$

where $\Delta H = H_B - H_A$ and $\Delta S = S_B - S_A$. It should be noted that both ΔH and ΔS are temperature-dependent according to respectively the Kirchhoff law (Chap. 1, Eq. (1.13)) and Eq. (2.12), cf. Chap. 2. A process is spontaneous if $\Delta G < 0$.

Fig. 3.2 The temperature
dependence of spontaneity
of a process depending on
the values (signs) of ΔH and
ΔS: green indicates
conditions for a spontaneous
process and red indicates
conditions for a
non-spontaneous process

	$\Delta H < 0$	$\Delta H > 0$
$\Delta S < 0$	$\Delta G < 0$ at low T $\Delta G > 0$ at high T	$\Delta G > 0$ at all T
$\Delta S > 0$	$\Delta G < 0$ at all T	$\Delta G > 0$ at low T $\Delta G < 0$ at high T

Figure 3.2 shows a helpful scheme which demonstrates the impact of ΔH and ΔS on ΔG and spontaneity.

The melting of a crystal at constant temperature requires a flow of heat from the surroundings, i.e. $\Delta H_m > 0$. The entropy of melting is also positive, $\Delta S_m > 0$. Considering a relatively narrow temperature region, it is a tolerable approximation to assume that ΔH_m and ΔS_m are constant. Using this approximation, the entropy of melting (ΔS_m) is then:

$$\Delta H_m - T_m \Delta S_m = 0 \Rightarrow \Delta S_m = \frac{\Delta H_m}{T_m} \tag{3.10}$$

where T_m is the melting point. The first expression in Eq. (3.10) is obtained by considering that $\Delta G_m = 0$ at the melting point (T_m). The temperature dependence of ΔG_m, according to Eqs. (3.9) and (3.10), is given by:

$$\Delta G_m = \Delta H_m - T \cdot \left(\frac{\Delta H_m}{T_m} \right) = \Delta H_m \cdot \left(1 - \frac{T}{T_m} \right) \tag{3.11}$$

Figure 3.3 shows the variation of ΔG_m with temperature. Melting is spontaneous when $\Delta G_m < 0$, i.e. at $T > T_m$. Chapter 6 presents more information about phase transitions.

Fig. 3.3 The change in
Gibbs free energy (ΔG_m) on
melting as a function of
temperature assuming
constant values for ΔH_m and
ΔS_m. The melting point (T_m)
is the lowest temperature for
spontaneous melting

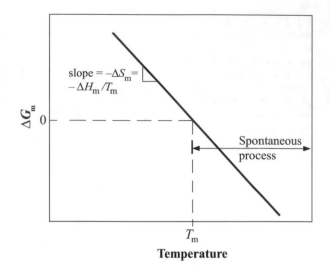

Chapter 4
A Comprehensive View of the State Functions Including Maxwell Relations

The fundamental state functions $(U, H, S, G, A, V, T, p$ and $n)$ are defined in Chaps. 1, 2, and 3. A set of useful relationships between the state functions was presented by James Clerk Maxwell, the great Scottish scientist, who became famous also for formulating the classical theory of electromagnetism in the well-known Maxwell equations. The *starting equation*, which is based on the first and second laws (cf. Chap. 2, Eq. (2.11)), is:

$$dU = TdS - pdV + \mu dn \therefore U = f(S, V, n) \qquad (4.1)$$

The total differential states that U is a function of entropy, volume and the number of moles of molecules. Equation (4.1) can be inserted in the total differentials derived from the *definitions of the other state function* as is here done for (a) the enthalpy (H):

$$H = U + pV \Rightarrow dH = dU + pdV + Vdp$$
$$dH = TdS - pdV + \mu dn + pdV + Vdp = TdS + Vdp + \mu dn \qquad (4.2)$$
$$H = f(S, p, n)$$

(b) the Helmholtz free energy (A):

$$A = U - TS \Rightarrow dA = dU - TdS - SdT$$
$$dA = TdS - pdV + \mu dn - TdS - SdT = -SdT - pdV + \mu dn \qquad (4.3)$$
$$A = f(T, V, n))$$

(c) and the Gibbs free energy (G):

© The Author(s), under exclusive license to Springer Nature Switzerland AG 2020
U. W. Gedde, *Essential Classical Thermodynamics*, SpringerBriefs in Physics,
https://doi.org/10.1007/978-3-030-38285-8_4

$$G = H - TS \Rightarrow dG = dH - TdS - SdT$$
$$dG = TdS + Vdp + \mu dn - TdS - SdT = -SdT + Vdp + \mu dn \qquad (4.4)$$
$$G = f(T, p, n)$$

Three relations are obtained immediately from the total differential of U:

$$dU = TdS - pdV + \mu dn. \therefore U = f(S, V, n)$$
$$dU = \left(\frac{\partial U}{\partial S}\right)_{V,n} dS + \left(\frac{\partial U}{\partial V}\right)_{S,n} dV + \left(\frac{\partial U}{\partial n}\right)_{S,V} dn \qquad (4.5)$$

And by comparison of the two expressions given in Eq. (4.5) that the three state functions (T, p and μ) are related to the partial derivatives, the following relations are derived:

$$T = \left(\frac{\partial U}{\partial S}\right)_{V,n} \wedge p = -\left(\frac{\partial U}{\partial V}\right)_{S,n} \wedge \mu = \left(\frac{\partial U}{\partial n}\right)_{S,V} \qquad (4.6)$$

The Maxwell relations are obtained by generating the *second derivatives* of the state functions in two possible ways. The order of derivation is unimportant (Schwarz theorem, named after the German mathematician Hermann Schwarz), as is shown for an example function $z = f(x,y)$ in Fig. 4.1a. Figure 4.1b shows the identification of state functions based on the first derivatives and Fig. 4.1c the method used to obtain the second derivative in two different ways. For each state function, three different Maxwell relations are available, which means that there are *12 Maxwell relations* for the four state functions.

Based on Eq. (4.6), application of the principle of the symmetry of the second derivative shows how one such equation is obtained:

$$\left(\frac{\left(\frac{\partial U}{\partial S}\right)_{V,n}}{\partial V}\right)_{S,n} = \left(\frac{\left(\frac{\partial U}{\partial V}\right)_{S,n}}{\partial S}\right)_{V,n} \Rightarrow \left(\frac{\partial T}{\partial V}\right)_{S,n} = -\left(\frac{\partial p}{\partial S}\right)_{V,n} \qquad (4.7)$$

and the other Maxwell relations connected with U are:

$$\left(\frac{\partial T}{\partial n}\right)_{S,V} = \left(\frac{\partial \mu}{\partial S}\right)_{V,n} \wedge -\left(\frac{\partial p}{\partial n}\right)_{S,V} = \left(\frac{\partial \mu}{\partial V}\right)_{S,n} \qquad (4.8)$$

and the Maxwell relations connected with the enthalpy (H) are:

Fig. 4.1 (a–c) Symmetry of
the second derivatives of a
state function

(a)

$$z = 3x^2 y + 2xy^2$$

$$\left(\frac{\partial z}{\partial x}\right)_y = 6xy + 2y^2 \quad \wedge \quad \left(\frac{\partial}{\partial y}\left(\frac{\partial z}{\partial x}\right)_y\right)_x = 6x + 4y$$

$$\left(\frac{\partial z}{\partial y}\right)_x = 3x^2 + 4xy \quad \wedge \quad \left(\frac{\partial}{\partial x}\left(\frac{\partial z}{\partial y}\right)_x\right)_y = 6x + 4y$$

(b)

$$dU = TdS - pdV + \mu dn \quad \therefore U = f(S,V,n)$$

$$dU = \left(\frac{\partial U}{\partial S}\right)_{V,n} dS + \left(\frac{\partial U}{\partial V}\right)_{S,n} dV + \left(\frac{\partial U}{\partial \mu}\right)_{S,V} d\mu$$

(c)

$$\left(\frac{\left(\frac{\partial U}{\partial S}\right)_{V,n}}{\partial V}\right)_{S,n} = \left(\frac{\left(\frac{\partial U}{\partial V}\right)_{S,n}}{\partial S}\right)_{V,n} \Rightarrow \left(\frac{\partial T}{\partial V}\right)_{S,n} = -\left(\frac{\partial p}{\partial S}\right)_{V,n}$$

$$H = f(S,p,n)$$

$$\left(\frac{\partial T}{\partial p}\right)_{S,n} = \left(\frac{\partial V}{\partial S}\right)_{p,n} \quad \wedge \quad \left(\frac{\partial T}{\partial n}\right)_{S,p} = \left(\frac{\partial \mu}{\partial S}\right)_{p,n} \quad \wedge \quad \left(\frac{\partial V}{\partial n}\right)_{S,p} = \left(\frac{\partial \mu}{\partial p}\right)_{S,n} \tag{4.9}$$

The equations obtained from the Helmholtz free energy (A) are:

$$A = f(T,V,n)$$

$$\left(\frac{\partial S}{\partial V}\right)_{T,n} = \left(\frac{\partial p}{\partial T}\right)_{V,n} \quad \wedge \quad \left(\frac{\partial S}{\partial n}\right)_{T,V} = -\left(\frac{\partial \mu}{\partial T}\right)_{V,n} \quad \wedge \quad \left(\frac{\partial p}{\partial n}\right)_{T,V} = -\left(\frac{\partial \mu}{\partial V}\right)_{T,n}$$

$$\tag{4.10}$$

and the equations obtained from the Gibbs free energy (G) are:

$$G = f(T, p, n)$$

$$\left(\frac{\partial S}{\partial p}\right)_{T,n} = -\left(\frac{\partial V}{\partial T}\right)_{p,n} \wedge \left(\frac{\partial S}{\partial n}\right)_{T,p} = -\left(\frac{\partial \mu}{\partial T}\right)_{p,n} \wedge \left(\frac{\partial V}{\partial n}\right)_{T,p} = \left(\frac{\partial \mu}{\partial p}\right)_{T,n}$$

(4.11)

The Maxwell relations constitute a *toolbox* which can be used to solve a range of different thermodynamic problems. If you reach a dead-end in a derivation, it may well be that one of the Maxwell relations can assist in *solving the problem*.

Chapter 5
Chemical Potential and Partial Molar Properties

The *chemical potential* belongs to the group of *partial molar properties*, and is useful for evaluating miscibility, and the effects on phase transitions (melting and evaporation) and osmotic pressure caused by the composition of solutions. The symbol of chemical potential is μ and the unit is joule per mole, J mol^{-1}, which demonstrates that μ is an *intensive property* (i.e. a property that is not dependent on the system size). The chemical potential can be defined with respect to G; the chemical potential μ_x is the partial derivative of G with respect to the number of moles of compound x (n_x) in a solution based on compound x and other compounds while keeping constant T and p and the number of moles of the other compounds.

Figure 5.1 shows two different binary systems. The insertion of an A-molecule into the two solutions has different impacts on the local potential energies depending on the compositions of the solutions. In the case of a pristine B system (Fig. 5.1a), the A-molecule only interacts with B-molecules, whereas in the other system, the inserted A-molecule interacts with both A- and B-molecules (Fig. 5.1b). This shows that the chemical potential of A (denoted μ_A) is composition dependent. The chemical potentials of the two compounds (μ_A and μ_B) in a binary solution can be defined in terms of the partial derivatives of G according to:

$$\mu_A = \left(\frac{\partial G}{\partial n_A}\right)_{T,p,n_B} \wedge \mu_B = \left(\frac{\partial G}{\partial n_B}\right)_{T,p,n_A} \tag{5.1}$$

The total differential of the Gibbs free energy (dG) is thus given (cf. Chap. 4; Eq. (4.4)) by:

$$dG = -SdT + Vdp + \mu_A dn_A + \mu_B dn_B \tag{5.2}$$

which means that

© The Author(s), under exclusive license to Springer Nature Switzerland AG 2020
U. W. Gedde, *Essential Classical Thermodynamics*, SpringerBriefs in Physics,
https://doi.org/10.1007/978-3-030-38285-8_5

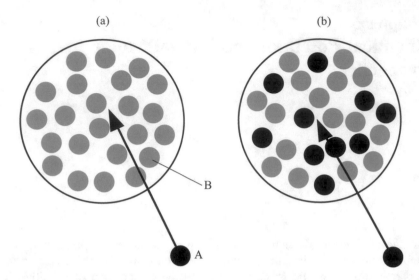

Fig. 5.1 Insertion of an A-molecule in (**a**) pristine B; (**b**) in a solution of A and B

$$G = f(T, p, n_A, n_B) \tag{5.3}$$

The chemical potential of A (μ_A) can also be defined from the other state functions according to Chap. 4, Eqs. (4.1, 4.2, and 4.3):

$$\mu_A = \left(\frac{\partial U}{\partial n_A}\right)_{S,V,n_B} \wedge \mu_A = \left(\frac{\partial H}{\partial n_A}\right)_{S,p,n_B} \wedge \mu_A = \left(\frac{\partial A}{\partial n_A}\right)_{T,V,n_B} \tag{5.4}$$

which shows that:

$$U = f(S, V, n_A, n_B) \wedge H = f(S, p, n_A, n_B) \wedge A = f(T, V, n_A, n_B) \tag{5.5}$$

At constant T and constant p, the following expression obeys:

$$G = n_A\mu_A + n_B\mu_B \tag{5.6}$$

The chemical potentials μ_A and μ_B are dependent on the composition and on p and T. Equation (5.6) can be differentiated at constant T and constant p to yield:

$$dG = \mu_A dn_A + \mu_B dn_B + n_A d\mu_A + n_B d\mu_B \tag{5.7}$$

The following expressions are obtained by comparing Eqs. (5.2) and (5.7):

$$\begin{aligned} n_A d\mu_A + n_B d\mu_B &= 0 \text{ (constant } p \text{ and } T) \\ n_A d\mu_A + n_B d\mu_B &= -SdT + Vdp \end{aligned} \tag{5.8}$$

These expressions are referred to as the variants of the *Gibbs-Duhem equation* (named after Willard Gibbs and the French physicist Pierre Duhem) for a binary solution. More general expressions, also referred to as the Gibbs-Duhem equations, are:

$$\sum_i n_i d\mu_i = 0 \text{ (constant } p \text{ and } T)$$

$$\sum_i n_i d\mu_i = -SdT + Vdp \tag{5.9}$$

The chemical potential of a pure compound (μ^*) is dependent on both T and p according to:

$$dG_m = -S_m dT + V_m dp$$

$$\mu^* = \left(\frac{\partial G}{\partial n}\right) = G_m \tag{5.10}$$

$$d\mu^* = -S_m dT + V_m dp$$

where G_m is the molar Gibbs free energy, which is equal to the chemical potential of the pure compound (μ^*; *note the asterisk* which marks that it deals with the pristine compound). The temperature dependence of μ^* is $(\partial\mu^*/\partial T)_p = -S_m$, which is always negative. The pressure dependence of μ^* is $(\partial\mu^*/\partial p)_T = V_m$, which is always positive. This topic is discussed further in Chap. 6.

The molar free energy of a binary solution (G_m), at a given p and T, is:

$$G_m = x_A \mu_A + x_B \mu_B$$

$$x_A + x_B = 1 \tag{5.11}$$

$$G_m = \mu_A + x_B(\mu_B - \mu_A)$$

The last expression in Eq. (5.11) can be differentiated considering its dependence on T and p:

$$dG_m = -S_m dT + V_m dp + (\mu_A - \mu_B)dx_B \Rightarrow \left(\frac{\partial G_m}{\partial x_B}\right)_{T,p} = \mu_A - \mu_B \tag{5.12}$$

The following equations (valid at a given combination of T and p) are obtained by inserting Eq. (5.12) in Eq. (5.11):

$$G_m = \mu_A + x_B(\mu_B - \mu_A) = \mu_A + x_B\left(\frac{\partial G_m}{\partial x_B}\right)$$

$$\mu_A = G_m - x_B\left(\frac{\partial G_m}{\partial x_B}\right) \tag{5.13}$$

A "symmetric" equation is obtained for μ_B:

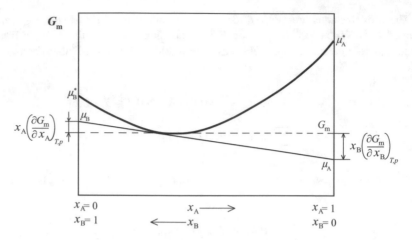

Fig. 5.2 Graphic demonstration of method to assess μ_A and μ_B based on Eqs. (5.13) and (5.14). The chemical potentials for the pure compounds are denoted $\mu_A{}^*$ and $\mu_B{}^*$

$$\mu_B = G_m - x_A \left(\frac{\partial G_m}{\partial x_A} \right) \tag{5.14}$$

Equations (5.13) and (5.14) provide the means for a method to assess the compositional dependence of μ_A and μ_B (Fig. 5.2).

This tool can be applied to assess the miscibility of compounds as a function of T, p and composition, provided that a *model* is available that expresses G_m as a function of the aforementioned parameters (cf. Chap. 7).

Apart from those based on G, the *partial molar quantities* include also a number of other state functions, e.g. V, H and S. Let us consider a binary system. The partial molar quantities are commonly considered at specific values of T and p:

$$\overline{V}_A = \left(\frac{\partial V}{\partial n_A} \right)_{T,p,n_B} \quad \overline{V}_B = \left(\frac{\partial V}{\partial n_B} \right)_{T,p,n_A}$$

$$\overline{H}_A = \left(\frac{\partial H}{\partial n_A} \right)_{T,p,n_B} \quad \overline{H}_B = \left(\frac{\partial H}{\partial n_B} \right)_{T,p,n_A} \tag{5.15}$$

$$\overline{S}_A = \left(\frac{\partial S}{\partial n_A} \right)_{T,p,n_B} \quad \overline{S}_B = \left(\frac{\partial S}{\partial n_A} \right)_{T,p,n_A}$$

where the quantities with bars above are the partial molar quantities.

Let us select the volume quantities for a demonstration. The volume of a system (V) at a specific T and p is given by:

$$V = n_A \overline{V}_A + n_B \overline{V}_B \tag{5.16}$$

Fig. 5.3 Molar quantity of
a generic state function (X_m)
of a solution based on two
components (A and B)
plotted as a function of the
molar ratio of component A
(x_A). The partial molar
values at $x_A = 0.6$ are shown

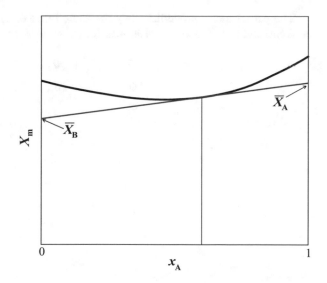

And for one mole of molecules ($n = n_A + n_B = 1$), the following expression holds:

$$\frac{V}{n} = \frac{n_A \overline{V}_A}{n} + \frac{n_B \overline{V}_B}{n} \Rightarrow V_m = x_A \overline{V}_A + x_B \overline{V}_B \tag{5.17}$$

where $V_m = V/n$ is the molar volume of the solution. As in the case of μ, the Gibbs-Duhem equation for the volume (cf. Eq. (5.8)) enables the assessment of the partial molar volumes of components A and B. All partial molar quantities can be determined according to the principle shown in Fig. 5.3, *a universal method!* In a plot of the molar property (X_m) versus the molar fraction, the values of the partial molar properties of each of the components are obtained as the intercepts at the y-axis of the tangents to the curves at the given compositions. The partial molar quantities of the *pure components* are obtained as the X_m values at $x_A = 0$ and 1.

An illuminating example is the volume of solutions of *ethanol* and *water*. For a chemist, not dealing with nuclear reactions, the mass is an invariant quantity, by combining 1 g A and 1 g B, a mixture weighing 2 g is obtained. Volumes of matter can in some cases be approximately additive but they need not to be so. A solution of water and ethanol mixed at 25 °C and 1 atm is one such example. Adding one mole water to pure water increases the system volume by 18 $(cm)^3$, whereas adding one mole water to pristine ethanol increases the system volume by only 14 $(cm)^3$. It is said that the famous Russian chemist *Dmitri Mendeleev* saved the neck of a person responsible for a Vodka factory in Russia, who was accused of theft. Mixing x L water with x L of 80% ethanol solution results in less than $2x$ L 40% vodka.

The partial molar properties obey the same types of expression as those of the conventional state functions, as shown, for example, for component A:

$$\overline{H}_A = \overline{U}_A + p\overline{V}_A$$
$$\overline{G}_A = \overline{H}_A + T\overline{S}_A \tag{5.18}$$

etc.

Chapter 6
One-Component Systems: Transitions and Phase Diagrams

This chapter deals with phases and phase transitions of *pure compounds*. A *phase* is a physically distinct form of matter, either of a pure substance or of a mixture of several substances. Examples of such phases are solids, liquids and gases. There are several possible solid phases for a given substance: different crystal phases (polymorphism) and glass. Liquid phases can in the case of mixtures coexist in different compositional forms. Thermodynamics deals with equilibrium states, and the *phase diagrams* that will be dealt with are obtained from such considerations. A phase diagram presents a convenient map showing where (i.e. at which T, p and composition) the different phases are thermodynamically stable. For pure substances, the composition is not an issue. The real structures formed may not however match the phase diagram; the *kinetics* often play a decisive role. However, the complexity of the kinetics is barely touched upon in the current text

A phase transition from gas to liquid (*condensation*) is illustrated in Fig. 6.1, a *phase diagram* which shows the presence of different phases in $p - V$ space. A third state function is also present, temperature (T). The lines in the pV-diagram indicate isotherms.

Let us follow the isotherm (at T_1) in the left-hand graph. The pressure is gradually increased by applying a force on the piston. The pressured volume is only a gas at point A. The ideal gas law, $pV = nRT$, at constant T confirms that $p \propto V^{-1}$, which is according to the path shown in the right-hand corner, also including A→B. At point B, small drops of a liquid phase are formed. At this combination of temperature (T_1) and pressure (p_1), the two phases (liquid and gas) are equally stable. More liquid is gradually formed (note the drawings associated with points C and D) by pushing the piston so that the volume decreases. This is done at constant pressure (p_1) and constant temperature (T_1). Condensation is exothermal, which means that heat has to be transferred to the surroundings in order to maintain constant temperature (cf. Chap. 1). At point E, only liquid is present, and further compression requires a substantial increase in pressure, since a liquid is much less compressible than a gas.

The red isotherm (marked T_c) shows the gradual change of the gas to a liquid-like medium referred to as a *supercritical fluid*. The point at the maximum of the

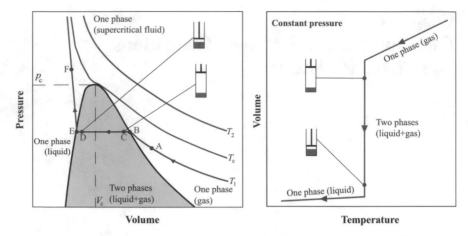

Fig. 6.1 Condensation of a gas shown in two diagrams: isothermal condensation (left) and isobaric condensation (right)

two-phase region (marked in grey) is the *critical point* (T_c, p_c and V_c). The right-hand graph shows condensation at constant pressure. When the temperature is gradually decreased at which the two phases have equal stability is reached and by very gentle pushing of the piston, the vertical blue line is followed until all the gas has been converted to liquid. Both graphs indicate the characteristic, stepwise decrease in volume associated with the phase transition (ΔV_{tr}). This is typical of several phase transitions, condensation, evaporation, crystallization and melting, which are referred to as *first-order transitions*. Let us now use the thermodynamics to derive relations between temperature and pressure for the first-order phase transitions. For a pure substance, the molar Gibbs free energy (G_m) is equal to the chemical potential (μ^*) (cf. Chap. 4, Eq. (4.4) and Chap. 5, Eq. (5.10)):

$$dG_m = -S_m dT + V_m dp \Leftrightarrow d\mu^* = -S_m dT + V_m dp \qquad (6.1)$$

where S_m and V_m are respectively the molar entropy and the molar volume. The partial derivatives of the chemical potential of a pure substance with respect to T and p are:

$$\left(\frac{\partial \mu^*}{\partial T}\right)_p = -S_m$$

$$\left(\frac{\partial \mu^*}{\partial p}\right)_T = V_m \qquad (6.2)$$

The condition for equilibrium between two phases of a substance, e.g. crystal (solid, s) and melt (liquid, l), is:

Fig. 6.2 Chemical potential as a function of temperature at constant pressure (isobars) of the three phases (crystal (blue), liquid (green) and gas (red)) of a pure substance. The black line indicates the equilibrium path with stable crystal ($T < T_\mathrm{m}$), stable liquid ($T_\mathrm{m} < T < T_\mathrm{b}$) and stable gas ($T > T_\mathrm{b}$)

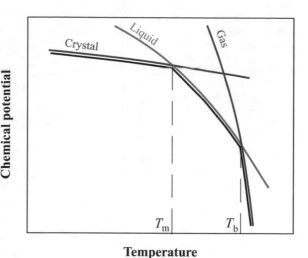

Temperature

$$\mu_\mathrm{s}^* = \mu_\mathrm{l}^* \Leftrightarrow d\mu_\mathrm{s}^* = d\mu_\mathrm{l}^* \tag{6.3}$$

Figure 6.2 shows the $\mu = f(T)$ isobars for the three phases: crystal, melt (liquid) and gas. The slope of each line is equal to $-S_\mathrm{m}$. Note that the molar entropy for a given phase shows a moderate increase with increasing temperature, i.e. the lines show curvature. The intersections of the curves occur at the melting temperature (blue and green curves, T_m) and the boiling temperature (green and red curves, T_b), respectively. At these temperatures, two phases have the same chemical potential. A reversible path for a phase transition, e.g. crystal melting, must go through the phase change at T_m.

The conventional pressure dependence of the melting temperature is illustrated in Fig. 6.3. The increase in chemical potential of each of the phases induced by an increase of pressure is $\Delta p V_\mathrm{m}$, which is greater for the liquid phase than for the crystal phase.

The melting of ice is however different from the conventional behaviour. The increase in chemical potential of the liquid phase (water) is less than that of the crystal phase (ice). Ice floats on water! Consequently, the melting point of ice decreases with increasing pressure.

The pressure dependence of the phase transition crystal (solid, s) → melt (liquid, l) can, based on Eqs. (6.1) and (6.3), be expressed as:

$$-S_\mathrm{m,s}dT + V_\mathrm{m,s}dp = -S_\mathrm{m,l}dT + V_\mathrm{m,l}dp \tag{6.4}$$

where the indices s and l refer to solid and liquid phases, respectively. This expression yields the *Clapeyron equation* (named after the French physicist Benoît Paul Émile Clapeyron, one of the pioneers of thermodynamics):

Fig. 6.3 Conventional pressure dependence of the melting temperature. The greater molar volume of the liquid compared to that of the crystal phase causes a greater pressure-induced displacement of the liquid line

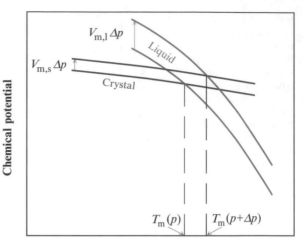

$$\frac{dp}{dT} = \frac{S_{m,l} - S_{m,s}}{V_{m,l} - V_{m,s}} = \frac{\Delta S_{m,tr}}{\Delta V_{m,tr}} \tag{6.5}$$

where $\Delta S_{m,tr}$ and $\Delta V_{m,tr}$ are respectively the molar entropy change and the molar volume change of the phase transition. For crystal melting, both $\Delta S_{m,tr}$ and $\Delta V_{m,tr}$ are usually positive, i.e. $dp/dT > 0$.

Only a few compounds show a different behaviour, but the *phase transition ice → water* is one exceptional case: $\Delta S_{m,tr} > 0$ (all compounds show this behaviour) but $\Delta V_{m,tr} < 0$ (very rare; the density of ice is less than that of water at 0 °C). In this particular case, $dp/dT < 0$. The transition liquid → gas is always accompanied by $dp/dT > 0$.

Equation (6.5) can be expressed differently by considering that the change in entropy of the transition (note that this is demonstrated for a melting transition) can be expressed as:

$$\Delta G_{m,m} = \Delta H_{m,m} - T\Delta S_{m,m} = 0 \Rightarrow \Delta S_{m,m} = \frac{\Delta H_{m,m}}{T}$$
$$dp = \frac{\Delta H_{m,m}}{\Delta V_{m,m}} \cdot \frac{dT}{T} \tag{6.6}$$

Equation (6.6) is integrated between a reference state (p^* and T^*) and an arbitrary state (p, T):

$$\int_{p*}^{p} dp = \frac{\Delta H_{m,m}}{\Delta V_{m,m}} \cdot \int_{T*}^{T} \frac{dT}{T} \Rightarrow p = p* + \frac{\Delta H_{m,m}}{\Delta V_{m,m}} \cdot \ln\left(\frac{T}{T*}\right) \qquad (6.7)$$

where it is assumed that the change in molar enthalpy of melting ($\Delta H_{m,m}$) and the change in molar volume of melting ($\Delta V_{m,m}$) are independent of temperature. This assumption is a good approximation considering a narrow temperature range. Without expressing the reference values, Eq. (6.7) can be written in a more general form according to:

$$p = \frac{\Delta H_{m,m}}{\Delta V_{m,m}} \cdot \ln T + C \qquad (6.8)$$

where C is a constant. The Clapeyron equation is applicable to all first-order transitions, a concept which is explained elsewhere in this chapter. However, Eq. (6.6) can be simplified for evaporation, when the gas obeys the ideal gas law, by first assuming that the change in molar volume accompanying evaporation, $\Delta V_{m,v}$, is approximately equal to the molar volume of solely the gas ($V_{m,g}$):

$$dp = \frac{\Delta H_{m,v}}{\Delta V_{m,v}} \cdot \frac{dT}{T} \approx \frac{\Delta H_{m,v}}{V_{m,g}} \cdot \frac{dT}{T} \approx \frac{\Delta H_{m,v}}{(RT/p)} \cdot \frac{dT}{T}$$
$$\frac{dp}{p} = \frac{\Delta H_{m,v}}{RT^2} dT \Rightarrow \frac{d\ln p}{dT} = \frac{\Delta H_{m,v}}{RT^2} \qquad (6.9)$$

Equation (6.9) is the *Clausius-Clapeyron equation*. Another form of the Clausius-Clapeyron equation is displayed in Eq. (6.10). It may be noted that the Clausius-Clapeyron equation shows that the derivative dp/dT is always positive, because $\Delta H_{m,v}$ is always positive:

$$\frac{d\ln p}{dT} = \frac{\Delta H_{m,v}}{RT^2} \Rightarrow \ln(p/p_0) = -\frac{\Delta H_{m,v}}{RT} + \frac{\Delta H_{m,v}}{RT_0}$$
$$\ln p = -\frac{\Delta H_{m,v}}{RT} + C \Leftrightarrow \frac{d\ln p}{d\left(\frac{1}{T}\right)} = -\frac{\Delta H_{m,v}}{R} \qquad (6.10)$$

Figure 6.4 shows a generic phase diagram of a pure compound. Three equilibria are displayed: solid $\rightarrow$ gas (below the triple point, described by the Clausius-Clapeyron equation), solid $\rightarrow$ liquid (above the triple point to the left, described by the Clapeyron equation) and liquid $\rightarrow$ gas (above the triple point to the right, described by the Clausius-Clapeyron equation at moderate pressures). Thus, each of the three lines defines the coexistence of two phases. The triple point indicates T and

Fig. 6.4 Phase diagram for a single component (pristine compound). The three phases are displayed and each line indicates coexistence of two phases. The triple point (three phases coexist) and the critical point are shown. The lines obey the Clapeyron (C) equation or the Clausius-Clapeyron (CC) equation

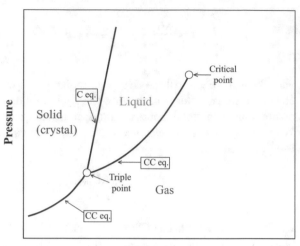

Temperature

p for the coexistence of three phases. Gibbs formulated a law that quantifies the degree of freedom (F, the number of intensive properties (i.e. a property that is not dependent on the system size, e.g. temperature, pressure or molar ratio)) that can be changed without changing the number of phases). *Gibbs phase rule* for a single component is:

$$F = C - P + 2$$
$$C = 1 \Rightarrow F = 3 - P \tag{6.11}$$

where C is the number of components (one in this case) and P is the number of phases.

Let us apply the Gibbs phase rule to the phase diagram shown in Fig. 6.4. In the areas outside the lines, only one phase is present, viz. $P = 1$ and thus $F = 2$, which means that both temperature and pressure can be changed without changing the number of phases. At the triple point, three phases coexist, i.e. $P = 3$ and $F = 0$, which means that no parameter can be changed without compromising the number of the existing phases. At the lines, two phases coexist, i.e. $P = 2$ and $F = 1$. The number of phases is two as long as the track follows the line, which means that both pressure and temperature change, but that one of them depends upon the other, $p = f(T)$, which reduces F from 2 to 1. Equation (6.11) also shows that the *maximum number of coexisting phases* in a pure compound is *three*.

The phase diagram of water is special; the equilibrium line between solid and liquid has a negative slope, $dp/dT < 0$. It has been suggested that the high pressure established between a skating metal blade and the ice during skating increases the pressure on the ice and decreases the melting point to an extent that a thin ice layer melts, which reduces the friction between the metal blade and the ice, but this is only

Fig. 6.5 Water content in vapour phase and partial water pressure as a function of temperature. The line indicates the equilibrium partial water pressure ($p_{H2O}{}^*$). The equilibrium partial pressure at 100 °C is 1 atm = 100 kPa

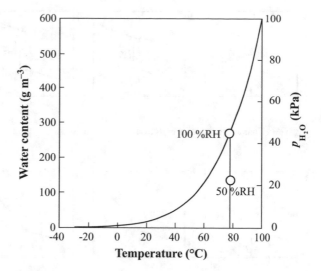

partly true. A very thin layer of water is covering the ice surface even at normal pressure and at temperatures clearly below the melting point.

The phase transition liquid water ↔ water vapour is of considerable practical importance. Figure 6.5 shows the pronounced increase in equilibrium water content with increasing temperature reaching 600 g per m³ gas volume at the boiling point (100 °C). The equilibrium partial pressure of water vapour at 100 °C is 1 atm ≈ 100 kPa. The water content at room temperature is only 20 g m^{-3}, which is less than 5% of that at 100 °C. The relative humidity (abbreviated RH) is defined with respect to the equilibrium partial water vapour pressure ($p_{H2O}{}^*$) according to %RH = $100 \cdot p_{H2O}/p_{H2O}{}^*$, where p_{H2O} is the actual partial water vapour pressure (cf. Fig. 6.5). Why is the indoor RH very low during winter in countries with cold climate? Assume that the outdoor temperature is −30 °C. At this temperature, the equilibrium water content is only 0.5 g m^{-3}, which is a very small fraction of the equilibrium water content at room temperature (20 g m^{-3}). Thus, the air which is taken into your home contains very little moisture, and this means that the relative humidity of this air when it is heated to 23 °C is very low indeed.

The last topic of this chapter is concerned with classification of phase transitions. The phase transitions concerned with pure substances are divided into two groups, according to the classification system proposed by the Austrian theoretical physicist Paul Ehrenfest. Figure 6.6 shows the characteristics of the well-known first-order transitions: melting, crystallization, evaporation and condensation. The primary state functions V, H and S show a *stepwise change* at the transition temperature, i.e. $\Delta V = V_\beta - V_\alpha$, $\Delta H = H_\beta - H_\alpha$ and $\Delta S = S_\beta - S_\alpha$; α and β being the two phases. This is natural to us, because we know that a liquid and a gas have very different molar volumes. The interaction between molecules in a liquid is different from that in a gas. In fact, the molecules in an ideal gas show no interaction ($\pi_T = 0$).

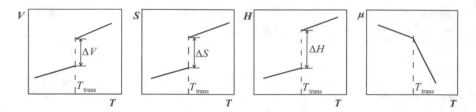

Fig. 6.6 Characteristics of a first-order phase transition

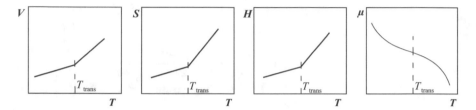

Fig. 6.7 Characteristics of second-order phase transition

The Gibbs free energy (chemical potential) is continuous, but its first temperature derivative is changing in a stepwise fashion. At the transition point, the Gibbs free energy of the two phases is equal, $G_\alpha = G_\beta$.

Second-order transitions or as they are also called, continuous phase transitions, are less well-known. The primary state functions (V, H and S) are continuous through the transition, but their first temperature derivatives (e.g. $\Delta \alpha$ and ΔC_p) show a stepwise change (Fig. 6.7). Examples of second-order transitions are *ferro-magnetic transitions*, *superconductive transitions* and *glass transitions*. Glasses show the characteristics of a second-order transition but the origin is in fact kinetic. However, there is an underlying thermodynamic second-order glass transition that is never observed in direct experiments. Note that the second derivative of the Gibbs free energy changes the sign at the transition point of a second-order transition (Fig. 6.7).

Phase transitions are conveniently studied by differential scanning calorimetry (DSC). Three transitions are observed in the thermogram displayed in Fig. 6.8. Two of the transitions are of first-order type, crystallization and melting. *First-order transitions* appear as peaks; in this diagram the endothermal process shows a peak pointing upwards. The area under the peaks (marked in the diagram) is proportional to the enthalpy change of the process. The area under the melting peak (proportional to the enthalpy of melting, ΔH_m) is only marginally greater than the area under the crystallization peak (proportional to the enthalpy of crystallization, ΔH_c), which indicates that the *degree of crystallinity* before the heating was small, the latter being

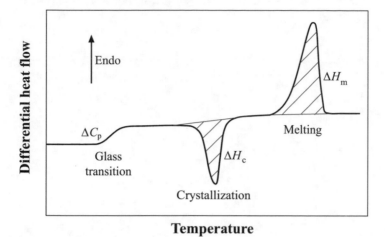

Fig. 6.8 Heating thermogram showing three phase transitions in poly(ethylene terephthalate): a glass transition with a jump in C_p (ΔC_p, apparent second-order transition), crystallization (first-order transition) with an exothermal peak (ΔH_c) and, at the highest temperatures, a melting transition (first-order transition) with an endothermal peak (ΔH_m)

$(\Delta H_m - \Delta H_c)/\Delta H_c^1$, where ΔH_c^1 is the enthalpy of melting of 100% crystalline polymer. The glass transition is indicated by a stepwise change (ΔC_p), and this is typical of an *apparent second-order transition*. The midpoint of the transition is referred to as the *glass transition temperature* (T_g).

Chapter 7
Solutions, Phase-Separated Systems, Colligative Properties and Phase Diagrams

The chemical potential of a substance A in a solution can be calculated by focusing on the gas phase above the solution. At equilibrium, the chemical potential of A in the solution is equal to the chemical potential of A in the gas phase, so that the first task is to calculate the chemical potential of A in the gas phase.

The chemical potential μ is dependent on the pressure (p) at constant temperature according to Eq. (6.2), i.e.:

$$\left(\frac{\partial \mu}{\partial p}\right)_T = V_{\mathrm{m}} \tag{7.1}$$

For a gas which obeys the ideal gas law, the following $\mu - p$ relationship is obtained:

$$d\mu = V_{\mathrm{m}} dp; V_{\mathrm{m}} = \frac{RT}{p} \Rightarrow d\mu = RT \frac{dp}{p}$$

$$\mu = \mu_0 + RT \ln\left(\frac{p}{p_0}\right) \tag{7.2}$$

where μ_0 is the chemical potential of the ideal gas at a reference pressure (p_0). Figure 7.1 shows (i) a closed system with a pure liquid A and a gas phase in equilibrium with the liquid phase (the pressure of component A is $p_A{}^*$) and (ii) a similar closed system but with a binary solution with the molar composition x_A. The chemical potential of the pure liquid A ($\mu_A{}^*$) is according to Eq. (7.2) given by:

$$\mu_A^* = \mu_{A,0} + RT \ln\left(\frac{p_A^*}{p_{A,0}}\right) \Rightarrow \mu_{A,0} = \mu_A^* - RT \ln\left(\frac{p_A^*}{p_{A,0}}\right) \tag{7.3}$$

and the chemical potential of A in the solution is given by:

Fig. 7.1 Sketch of a system consisting to the left of pure liquid phase A with a head-space gas phase with an equilibrium partial pressure p_A^*. To the right, the same for a solution of A with another substance (B). The equilibrium partial pressure of A is p_A

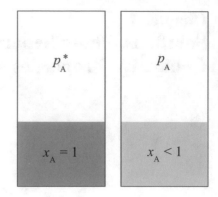

$$\mu_A = \mu_{A,0} + RT \ln \left(\frac{p_A}{p_{A,0}} \right) \tag{7.4}$$

The following expression is obtained by combining Eqs. (7.3) and (7.4):

$$\mu_A = \mu_A^* - RT \ln \left(\frac{p_A^*}{p_{A,0}} \right) + RT \ln \left(\frac{p_A}{p_{A,0}} \right) \Rightarrow \mu_A = \mu_A^* + RT \ln \left(\frac{p_A}{p_A^*} \right) \tag{7.5}$$

The French chemist Francois-Marie Raoult proposed the following relationship between the partial pressure of compound A in the gas phase and the molar composition (x_A) of the liquid solution:

$$p_A = x_A p_A^* \tag{7.6}$$

This relationship, which is referred to as *Raoult's law,* is valid for *ideal solutions.* The definition of an ideal solution is a solution which obeys Raoult's law!

Figure 7.2 shows the partial pressures and the total pressure as a function of x_A for binary solutions of A and B. The partial pressure A is proportional to the molar ratio of the substance in the solution (Eq. (7.6)). Inserting Raoult's law in Eq. (7.5) yields:

$$\mu_A = \mu_A^* + RT \ln x_A \tag{7.7}$$

Equation (7.7) is valid only for an ideal solution. In general, including nonideal solutions, the following relationship is obtained by defining the *activity* of compound A (a_A) and inserting this into Eq. (7.5):

$$a_A = \frac{p_A}{p_A^*} ; \mu_A = \mu_A^* + RT \ln a_A \tag{7.8}$$

The activity can be related to the molar composition according to:

$$a_A = \gamma_A x_A ; \mu_A = \mu_A^* + RT \ln x_A + RT \ln \gamma_A \tag{7.9}$$

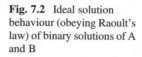

Fig. 7.2 Ideal solution behaviour (obeying Raoult's law) of binary solutions of A and B

(a)   **(b)**

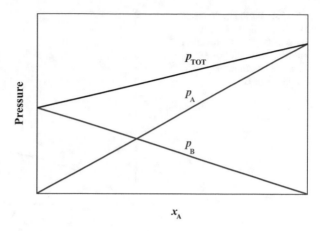

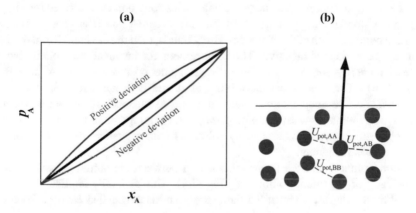

Fig. 7.3 (a) Deviation from ideal solution behaviour (the black line indicates ideal solution). (b) A sketch of the reason for the deviation

where γ_A is the *activity coefficient* of substance A. It should be noted that γ_A is a function of x_A. Figure 7.3 shows nonideal solution behaviour, either a positive deviation $p_A > p_A^* \cdot x_A$ or a negative deviation $p_A < p_A^* \cdot x_A$. In both cases, the reason is that the interaction between A and B ($U_{pot,AB}$) is different from that predicted by considering the A$\cdots$A ($U_{pot,AA}$) and B$\cdots$B ($U_{pot,BB}$) interactions:

$$U_{pot,AB} = \frac{U_{pot,AA} + U_{pot,BB}}{2} \quad \text{(ideal solution)}$$

$$U_{pot,AB} < \frac{U_{pot,AA} + U_{pot,BB}}{2} \quad \text{(negative deviation)} \qquad (7.10)$$

$$U_{pot,AB} > \frac{U_{pot,AA} + U_{pot,BB}}{2} \quad \text{(positive deviation)}$$

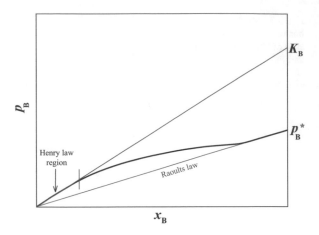

Fig. 7.4 Partial pressure of B as a function of the molar ratio of B (x_B) of a nonideal solution which follows Henry's law in dilute solutions of B

The potential energies take negative values; a strong bond is indicated by a highly negative potential energy value. If the bond between A and B is stronger than the average bond strength of the A⋯A and B⋯B bonds, then the deviation from the ideal solution behaviour is negative. The requirement for an ideal solution is that the potential energy for A⋯B is the average value of those of A⋯A and B⋯B (Eq. (7.10)). This means that the mixing of A and B in the ideal solution case is *athermal*, i.e. the temperature remains unchanged on mixing. Solutions having a negative deviation show *exothermal mixing*, i.e. the temperature increase on mixing. The opposite behaviour is shown by solutions with a positive deviation from the ideal solution behaviour.

Dilute solutions show a linear relationship between the partial pressure and the molar ratio of the minor component (Fig. 7.4). This generally applicable law was coined by the English chemist William Henry and is thus called *Henry's law*:

$$p_B = K_B x_B \tag{7.11}$$

where K_B is referred to as the Henry law constant. The idea behind this law is intuitively fairly simple to understand. The surroundings of a B molecule in a dilute solution are well-defined uniformly with A. Hence, the vapour pressure of B should be proportional to the number of available B molecules in the vicinity of the surface and thus proportional to the average concentration (x_B).

Let us calculate the molar Gibbs free energy of mixing two components (A and B), which form an ideal solution. The total free energy of the pristine components (G_0) is:

$$G_0 = n_A \mu_A^* + n_B \mu_B^* \tag{7.12}$$

According to Eq. (7.7), the free energy of the ideal solution is equal to:

Fig. 7.5 Sketch of the mean-field surroundings around an A- and a B-molecule. In this case the number of neighbours is six ($z = 6$), of which 1/3 are blue (B) and 2/3 are red (A)

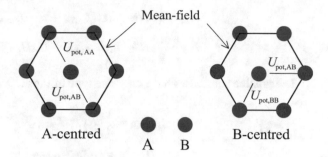

$$G = n_A \left(\mu_A^* + RT \ln x_A \right) + n_B \left(\mu_B^* + RT \ln x_B \right) \qquad (7.13)$$

The free energy of mixing (ΔG_{mix}) is the difference between G and G_0:

$$\Delta G_{mix} = G - G_0 = n_A \left(\mu_A^* + RT \ln x_A \right) + n_B \left(\mu_B^* + RT \ln x_B \right) - n_A \mu_A^* - n_B \mu_B^*$$
$$\Delta G_{mix} = RT(n_A \ln x_A + n_B \ln x_B)$$

$$(7.14)$$

which can be scaled to include a total of one mole of molecules (including both A and B):

$$\Delta G_{mix,m} = RT \left(\frac{n_A}{n} \ln x_A + \frac{n_B}{n} \ln x_B \right) = RT(x_A \ln x_A + x_B \ln x_B) \qquad (7.15)$$

The enthalpy of mixing for an ideal solution is athermal, i.e.:

$$\Delta H_{mix,m} = 0 \Rightarrow \Delta S_{mix,m} = -\frac{\Delta G_{mix,m}}{T}$$
$$\Delta S_{mix,m} = -R(x_A \ln x_A + x_B \ln x_B) \qquad (7.16)$$

Deviations from the ideal solution behaviour can be described by using a *mean-field approximation* to express ΔH_{mix} (Fig. 7.5). This method assumes that an average concentration of A and B molecules are present around each A-molecule. Two different interactions are possible in this configuration, A$\cdots$A (potential energy $= U_{pot,AA}$) and A$\cdots$B (potential energy $= U_{pot,AB}$); and the numbers of interactions are zx_A (A$\cdots$A) and zx_B (A$\cdots$B), respectively, where z is the coordination number, i.e. the number of neighbours around each molecule. The B-centred units have only B$\cdots$B and B$\cdots$A contacts. The enthalpies of the A-centred interactions (H_A) and B-centred interactions (H_B) are:

$$H_A = \frac{1}{2} \cdot Nx_A \left(x_A z U_{pot,AA} + x_B z U_{pot,AB} \right)$$

$$H_B = \frac{1}{2} \cdot Nx_B \left(x_B z U_{pot,BB} + x_A z U_{pot,AB} \right)$$

(7.17)

The pristine substances have only one type of interaction:

$$H_{A,0} = \frac{1}{2} \cdot Nx_A z U_{pot,AA}$$

$$H_{B,0} = \frac{1}{2} \cdot Nx_B z U_{pot,BB}$$

(7.18)

The enthalpy change associated with mixing is thus given by:

$$\Delta H_{mix} = (H_A + H_B) - (H_{A,0} + H_{B,0})$$

(7.19)

By combining Eqs. (7.17, 7.18, and 7.19), the following expression is obtained:

$$\Delta H_{mix} = Nz\Delta U_{pot,AB} x_A x_B$$

(7.20)

where

$$\Delta U_{pot,AB} = U_{pot,AB} - \frac{\left(U_{pot,AA} + U_{pot,BB} \right)}{2}$$

(7.21)

This equation expresses the difference between the potential energy of the A$\cdots$B contact and the average potential energy of the contacts between the pure substances (A$\cdots$A and B$\cdots$B), and $\Delta U_{pot,AB}$ is conveniently combined with the coordination number z to a single factor B that can take both negative and positive values. Thus, the molar mixing enthalpy $\Delta H_{mix,m}$ takes either negative or positive values depending on the sign of B.

$$\Delta H_{mix,m} = \frac{1}{n} \cdot nz\Delta U_{pot,AB} x_A x_B = Bx_A x_B$$

(7.22)

The product $x_A x_B$ expresses the number of contacts between A and B depending on the molar ratios. For a binary solution, the number of contacts (y) is $x_A \cdot (1 - x_A) = x_A - x_A^2 \Rightarrow \partial y/\partial x_A = 1 - 2\,x_A$; and the maximum number of contacts is obtained at $x_A = x_B = \frac{1}{2}$. This implies that $\Delta H_{mix,m}$ shows either a minimum ($B < 0$) or a maximum ($B > 0$) at $x_A = \frac{1}{2}$. The molar Gibbs free energy of mixing ($\Delta G_{mix,m}$) is the sum of the molar entropy of mixing and the molar enthalpy of mixing:

$$\Delta G_{mix,m} = RT(x_A \ln x_A + x_B \ln x_B) + Bx_A x_B$$

(7.23)

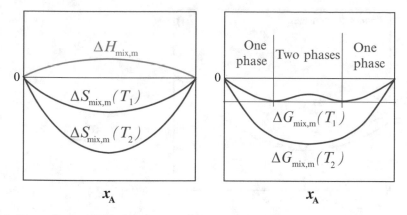

Fig. 7.6 Left: Molar enthalpy of mixing ($B > 0$) and molar entropy of mixing for two temperatures ($T_2 > T_1$) as a function of the molar ratio of A. Right: Molar free energy of mixing at two temperatures ($T_2 > T_1$) as a function of the molar ratio of A. At T_2: miscible system in the entire composition range. At T_1: a two-phase system surrounded by a single-phase system

The first term is always negative with a minimum at $x_A = \frac{1}{2}$, and it is symmetric about the minimum. The depth of the minimum increases with increasing temperature. The second term can take either a positive value (maximum at $x_A = \frac{1}{2}$) or a negative value (minimum at $x_A = \frac{1}{2}$). Figure 7.6 sketches the entropy contribution at two different temperatures (first term in Eq. (7.23); $T_2 > T_1$) and the enthalpy contribution (second term in Eq. (7.23)) which is independent of temperature in this simple case. The right-hand diagram shows the sum of the two contributions ($\Delta G_{mix,m}$) as a function of x_A. The high temperature curve shows a positive second derivative which indicates miscibility at all compositions at T_2. At T_1, however, between the minima (which can be connected by a single tangent), a two-phase system is the most stable. Outside these points (referred to as binodals), the substances are fully miscible (a single phase is present).

Figure 7.7a shows the molar free energy of a binary solution as a function of x_A. The intercepts ($x_A = 1$ and $x_B = 1$) are μ_A^* and μ_B^*, respectively. The second derivative ($\partial^2 G_m / \partial x_A^2$) is positive, which means that a single phase always has a lower free energy than any combination of two separate phases with the correct overall composition (Fig. 7.7a; the overall composition $x_{A,total}$ is the one linked to the circular point and the virtual phases are marked with the vertical lines, with compositions $x_{A,total} + \Delta x_A$ and $x_{A,total} - \Delta x_A$, respectively). The single solution is thus thermodynamically stable. Figure 7.7b shows an isotherm (G_m vs. x_A) for a solution with an intermediate maximum, similar to that displayed for $\Delta G_{mix,m}$ (T_1) in Fig. 7.6. The tangent to the two points on the curve defines the binodal points and the corresponding x_A-values ($x'_{A,B}$ and $x''_{A,B}$) indicate the compositions of the two phases (denoted prime and double prime) present in the two-phase region, i.e. for x_A between $x'_{A,B}$ and $x''_{A,B}$. The *condition for the equilibrium* between the two phases is fulfilled when:

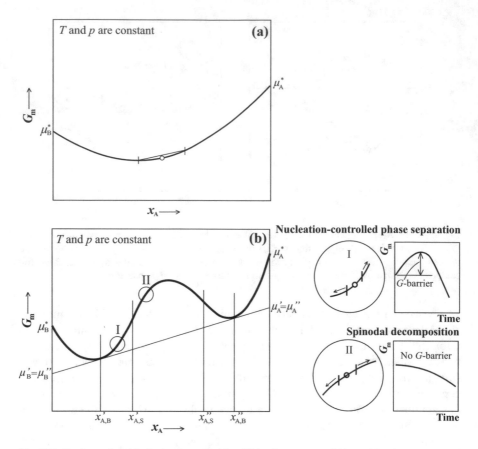

Fig. 7.7 Isothermal and isobaric plots of molar Gibbs free energy of binary blends composed of compounds A and B. From Gedde and Hedenqvist (2019)

$$\mu_A{}' = \mu_A{}'' \quad \wedge \quad \mu_B{}' = \mu_B{}'' \tag{7.24}$$

which is shown in Fig. 7.7b. Another important pair of points is obtained from the second derivative of G_m according to:

$$\frac{\partial^2 G_m}{\partial x_A^2} = 0 \tag{7.25}$$

The condition prescribed by Eq. (7.25) yields the points at which the curvature of $G_m - x_A$ changes from convex to concave or vice versa. This has an impact on the mechanism of phase separation (Fig. 7.7b, right-hand part):

(i) In the region where $\partial^2 G_m / \partial x_A^2 < 0$, phase separation occurs without a free energy barrier. This mechanism is called *spinodal decomposition*.

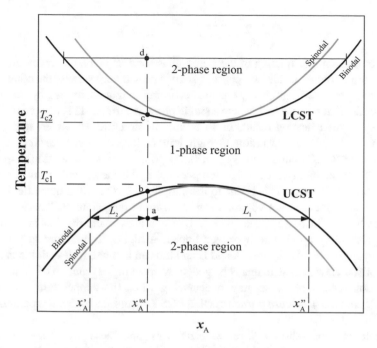

Fig. 7.8 Generic isobaric phase diagram for a binary liquid system

(ii) In the region where $\partial^2 G_m / \partial x_A^2 > 0$, *phase separation* occurs by a *nucleation* mechanism, viz. by a passage over a free energy barrier.

Figure 7.8 shows a generic phase diagram for a binary liquid system. The typical behaviour is that associated with the *upper critical solution temperature* (*UCST*). High temperatures favour solubility (miscibility), and this is described by the behaviour according to Eq. (7.23). The effect of temperature on the part linked to the entropy increase, viz. $T \cdot \Delta S_{mix,m}$, and the constant enthalpy contribution, $B x_A x_B$, is that the "hill" present in the central part of the $G - x_A$ isotherm due to a positive value of B decreases with increasing temperature and at a critical temperature (T_{c1}) of the UCST curve. Within the two-phase region, there are two different domains marking different phase separation mechanisms: (a) *nucleation-controlled phase separation* between the binodal and spinodal curves and (b) *spinodal decomposition* surrounded by the spinodal curve. The reading of the phase diagram can be demonstrated by taking a close look at point *a* in Fig. 7.8. Point *a* is in the spinodal decomposition region. At equilibrium, two phases are established with molar ratios x_A' and x_A''. These are obtained by drawing a horizontal line through point *a* and to give the intersections with the binodal curve. The amounts of the phases (denoted prime (') and double prime ('')) are obtained by applying the *lever rule*:

$$x' = \frac{L_1}{L_1 + L_2}; \quad x'' = \frac{L_2}{L_1 + L_2} \tag{7.26}$$

where x' is the molar fraction of the prime-phase and x'' is the molar fraction of the double-prime-phase. If this two-phase system is heated very slowly, the composition of the two phases will gradually change according to that prescribed by the binodal curve until point b is reached, where a single phase is formed. This may be the end of the story; further heating results in no change in structure. However, some binary systems show phase separation at high temperatures, a *lower critical solution temperature (LCST)* behaviour (Fig. 7.8). At point c, phase separation is initiated. The LCST behaviour (with a critical temperature T_{c2}) is incompatible with Eq. (7.23) and a constant B-value. This is due to the simplicity of the initial mean field model, which assumed a constant volume even at different temperatures. Thermal expansion leads to an increase in the distance between atoms, and this has an impact on strong secondary bonds like hydrogen bonds. A higher temperature may result in fewer hydrogen bonds between A and B and thus an increase in B with increasing T. A simplistic view is to state that B is, generally speaking, a function of T, and this is the reason for the phase separation occurring when the temperature is increased (LCST behaviour). A *word paradox* is that LCST occurs at a higher temperature than UCST, i.e. that $T_{c1} < T_{c2}$.

Dilute solutions show *colligative properties,* and these properties are simply proportional to the molar ratio of the solute. This word, which was introduced by Wilhelm Oswald in 1891, originates from Latin and means "bound together". The colligative properties are specifically the *osmotic pressure* (π), the *depression in vapor pressure*, the *depression in melting point* ($\Delta T_m = T_m - T$), and the *increase in the boiling point* at constant pressure ($\Delta T_b = T - T_b$). These quantities are related to the major component (the solvent denoted A), and the changes are caused by the inclusion of small amounts of solute (denoted B). The quantities concerned with pure A are T_m and T_b, whereas the transition temperatures concerned with the solution are labelled only T. The colligative properties (denoted CP) are a function of the molar ratio of the solute (x_B):

$$CP = K \cdot x_B \tag{7.27}$$

where K is a constant independent of x_B and also of the type of solute substance (B). The parameter K depends only on the characteristics of the major component (substance A, the solvent). The osmotic pressure is generated in a system-configuration as sketched in Fig. 7.9.

The *equilibrium condition* for the *osmotic configuration* is that the chemical potential of the pure solvent at pressure p is equal to the chemical potential of the solution with a molar fraction of component A of x_A and pressure $p + \pi$, viz.:

$$\mu_A^*(p) = \mu_A(x_A, p + \pi) \tag{7.28}$$

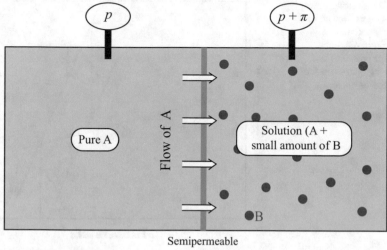

Fig. 7.9 *Osmotic pressure* (π) build-up by separating the pure solvent (left) from the solution (right) by a *semipermeable membrane*. This membrane allows only the transport of solvent (A) molecules. The solute (substance B) is unable to pass through the membrane

The chemical potential of the solution can be determined by considering the concentration dependence (Eq. (7.8)) and pressure dependence (Chap. 6, Eq. (6.2)):

$$\mu_A(x_A, p + \pi) = \mu_A^*(p + \pi) + RT \ln a_A =$$
$$\mu_A^*(p) + \int\limits_{p}^{p+\pi} V_{m,A}\, dp + RT \ln a_A \tag{7.29}$$

This integral can be solved by assuming that $V_{m,A}$ is constant, which is reasonable because only moderate changes in pressure are possible:

$$\mu_A(x_A, p + \pi) = \mu_A^*(p) + V_{m,A}\pi + RT \ln a_A \tag{7.30}$$

The following expression is obtained by combining Eqs. (7.28) and (7.30):

$$\mu_A^*(p) = \mu_A^*(p) + V_{m,A}\pi + RT \ln a_A \Rightarrow \pi = -\left(\frac{RT}{V_{m,A}}\right) \cdot \ln a_A \tag{7.31}$$

For an *ideal solution*, the activity a_A can be replaced by the molar ratio x_A:

Fig. 7.10 Melting point depression as viewed in a chemical potential-temperature diagram (constant pressure) and displaying quantities further explained in the text

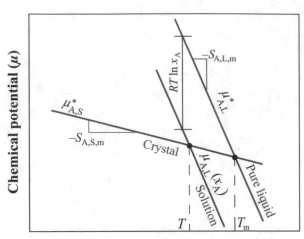

$$\pi = -\left(\frac{RT}{V_{m,A}}\right) \ln x_A \qquad (7.32)$$

and the relationship between the osmotic pressure and the concentration of solute (B) is obtained for a *dilute solution state* as:

$$\ln x_A = \ln (1 - x_B)$$
$$x_B \to 0 : \ln (1 - x_B) \approx -x_B \Rightarrow \ln x_A \approx -x_B \qquad (7.33)$$

which, after insertion in Eq. (7.32), gives:

$$\pi = \left(\frac{RT}{V_{m,A}}\right) \cdot x_B \qquad (7.34)$$

A unit-analysis based on Eq. (7.34) is $(J \cdot mol^{-1} \cdot K^{-1}) \cdot K/(m^3 \cdot mol^{-1}) = J \cdot m^{-3} = N \cdot m \cdot m^{-3} = N \cdot m^{-2} = Pa$; the osmotic pressure ($\pi$) is truly a pressure with the SI-unit of pascal, Pa. The structure of Eq. (7.34) is a good illustration of the colligative property concept; π is proportional to x_B, and the proportionality constant contains only quantities related to the solvent.

The next task is to demonstrate the colligative property character of the *melting point*. The condition for a pure crystal phase (substance A) in equilibrium with a solution consisting of A with molar ratio x_A is (cf. Fig. 7.10):

$$\mu_{A,S}^*(T) = \mu_{A,L}(x_A, T) \qquad (7.35)$$

where $\mu_{A,S}^*(T)$ is the chemical potential of the pure solid substance (A) at T and $\mu_{A,L}(x_A, T)$ is the chemical potential of substance A in the solution with molar ratio

x_A at T. The chemical potential of the solid phase at T can be related to its chemical potential at the conventional melting point (T_m), i.e. at which the crystalline and liquid phases are coexisting, according to Eq. (5.10):

$$\mu_{A,S}^*(T) = \mu_{A,S}^*(T_m) + \int_{T_m}^{T} (-S_{A,S,m})dT =$$
$$\mu_{A,S}^*(T_m) + S_{A,S,m}(T_m - T)$$

(7.36)

where $S_{A,S,m}$ is the molar entropy of solid substance A. It is assumed that $S_{A,S,m}$ is independent of temperature within the narrow temperature region considered. A similar calculation is carried out for the solution considering both a temperature shift from T_m to T and a concentration shift from the pure state to x_A:

$$\mu_{A,L}(x_A, T) = \mu_{A,L}^*(T) + RT \ln x_A =$$
$$\mu_{A,L}^*(T_m) + \int_{T_m}^{T} (-S_{A,L,m})dT + RT \ln x_A =$$
$$\mu_{A,L}^*(T_m) + S_{A,L,m}(T_m - T) + RT \ln x_A$$

(7.37)

where $S_{A,L,m}$ is the molar entropy of liquid A. The equilibrium equation for the pure state is:

$$\mu_{A,S}^*(T_m) = \mu_{A,L}^*(T_m)$$

(7.38)

Combining Eqs. (7.35, 7.36, and 7.37) yields:

$$\mu_{A,S}^*(T_m) + S_{A,S,m}(T_m - T) = \mu_{A,L}^*(T_m) + S_{A,L,m}(T_m - T) + RT \ln x_A$$ (7.39)

which, after insertion of Eq. (7.38), gives:

$$S_{A,S,m}(T_m - T) = S_{A,L,m}(T_m - T) + RT \ln x_A \Rightarrow$$
$$(S_{A,L,m} - S_{A,S,m})(T_m - T) = -RT \ln x_A$$

(7.40)

and finally, by expanding $\ln x_A = \ln(1 - x_B)$ according to Eq. (7.33):

$$(S_{A,L,m} - S_{A,S,m})(T_m - T) = RT \cdot x_B$$
$$T_m - T = \left(\frac{RT}{\Delta S_{A,m,m}}\right) \cdot x_B = \left(\frac{RT^2}{\Delta H_{A,m,m}}\right) \cdot x_B$$

(7.41)

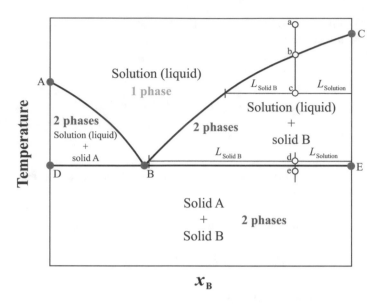

Fig. 7.11 Generic binary, isobaric phase diagram for substances that are completely miscible in the liquid state

where $\Delta S_{A,m,m}$ is the molar melting entropy of A and $\Delta H_{A,m,m}$ is the molar melting enthalpy of A. A similar relationship can be derived for the boiling point elevation $(T - T_b)$:

$$T - T_b = \left(\frac{RT^2}{\Delta H_{A,v,m}}\right) \cdot x_B \qquad (7.42)$$

Equations (7.41) and (7.42) both have the same general structure as the osmotic pressure equation (Eq. (7.34)); all three expressions obey the structure of Eq. (7.27). With this information, we can safely progress into the beautiful world of *binary phase diagrams*. Figure 7.11 shows a generic isobaric phase diagram based on two components which are fully miscible in the liquid state. Both components *crystallize separately*; cocrystallization is banned. The solution phase (phase 1) is placed in the upper part. Depending on the composition, crystallization starts with one of the two components. At compositions left of the *eutectic point*, crystallization of substance A occurs, whereas for the solutions with composition on the right-hand side of the eutectic point, crystallization of substance B occurs. The melting point for substance A is depressed due to the increase in concentration of the minor component B according to Eq. (7.41). Let us analyse the path from point *a* to *e* in the phase diagram shown in Fig. 7.11. Point *a* is in the liquid solution domain, where only one phase is present. After the solution is cooled and reaching point *b* is reached, a new phase, solid B, appears together with the solution phase. The amount of each of the two phases is calculated according to the *lever rule* (Eq. (7.26) which applied to

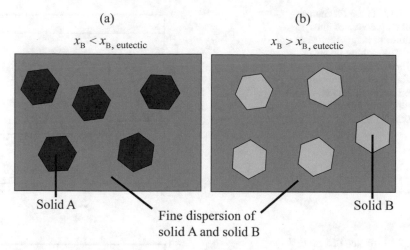

Fig. 7.12 Morphologies of structures obtained after cooling from the liquid solution state below (**a**) and above (**b**) the eutectic molar ratio of B

point b is $x_{\text{Solid B}} = L_{\text{Solid B}}/(L_{\text{Solid B}} + L_{\text{Solution}})$; $x_{\text{Solution}} = 1 - x_{\text{Solid B}}$. The chemical potential of substance B in the two phases is the same. During further cooling, reaching point c, further crystallization occurs generating more solid B, and, naturally, the concentration of substance B in the solution gradually decreases. Further cooling reaches point d, almost "touching" the minimum temperature for the solution phase and the eutectic composition, where two phases exist, solid B and a solution with the eutectic composition (ca. 30 mol% B in the current example). The amount of each phase can be calculated by applying the lever rule. On further cooling, reaching point e, the solution with the eutectic composition has crystallized into a fine-textured structure of pristine A and B crystals. This process is sometimes referred to as the *eutectic reaction*. The morphology consists of large solid B domains gradually formed during the path through the solution + solid B region and fine-textured biphasic parts from the eutectic reaction. Figure 7.12 shows the structures after cooling a solution with $x_B < x_B$ (eutecticum) and as in the displayed case (path $a - e$), a solution with $x_B > x_B$ (eutecticum). The resulting structures are clearly different, but in the phase diagram they are displayed merely as a two-phase system consisting of solid A and solid B. Each line A–B and C–B is referred to as a *liquidus lines*. The line DBE is called a *solidus line*. Point B is called the *eutectic point*. The horizontal lines connecting phases which are in equilibrium are called *tie lines*.

Figure 7.13 shows a slightly more complex phase diagram. Two eutectic points and four liquidus lines are present. It is as though the previous phase diagram (Fig. 7.11) has been doubled. This is almost true: an intermediate solid (crystalline) phase is present, Zn_2Mg. The melting points of the three crystalline phases in equilibrium with melts of the same composition are shown in the graph. This kind of melting is referred to as *congruent* melting.

Fig. 7.13 Phase diagram at constant pressure of binary mixtures of magnesium and zinc. The melting points of the three congruently melting species are shown. (Sketched from data of Friedrich and Mordike (2006))

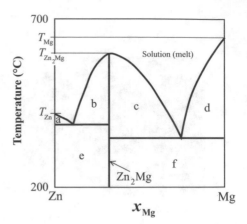

Fig. 7.14 Sketch of a phase diagram at constant pressure containing two peritectic points. (Drawn after data of Tollefsen et al. (2012))

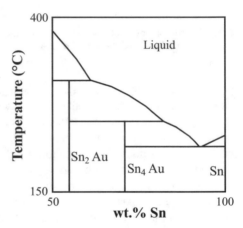

The following phases are present in the phase diagram shown in Fig. 7.13: (a) Zn (s) and solution; (b) solution and $Zn_2Mg(s)$; (c) $Zn_2Mg(s)$ and solution; (d) solution and Mg(s); (e) Zn(s) and $Zn_2Mg(s)$; and (f) $Zn_2Mg(s)$ and Mg(s).

Some compounds melt in a different way from the congruently melting species. When heated, such a compound decomposes into a new solid and a liquid (melt); both these phases have a molar composition different from that of the original compound. Such a compound shows *incongruent melting*, and this reaction is referred to as a *peritectic reaction*. A phase diagram showing this kind of behaviour is shown in Fig. 7.14. In fact, two of the intermediate compounds, Sn_2Au and Sn_4Au, show incongruent melting. The latter (Sn_4Au) decomposes into Sn_2Au and a molten phase. The point at which the peritectic reaction occurs is denoted the *peritectic point*.

A number of other liquid-solid phase diagrams have been reported, viz. (a) monotectic systems which in addition to containing eutectic and peritectic systems also show coexisting phases in the liquid state; (b) systems showing polymorphism, where several crystal phases exist; and (c) systems showing solid solution phases. For a more comprehensive treatment of this topic, the reader is referred to the specialized literature on phase diagrams.

The final part of this chapter is dedicated to binary phase diagrams involving *liquids* and *gases*. An ideal solution is chosen for simplicity, where the partial pressures (p_A and p_B) are according to Raoult's law:

$$p_A = x_A p_A^* \tag{7.43}$$

$$p_B = x_B p_B^* = (1 - x_A) p_B^* \tag{7.44}$$

where x_A and x_B are the molar ratios of A and B in the liquid and p_A^* and p_B^* are the partial pressures above the pure liquid substances. The total pressure is the sum of the two partial pressures:

$$p = p_A + p_B = x_A p_A^* + (1 - x_A) p_B^* = p_B^* + (p_A^* - p_B^*) x_A \tag{7.45}$$

The molar ratio of A in the gas phase (y_A) is given by:

$$y_A = \frac{p_A}{p} = \frac{x_A p_A^*}{p_B^* + (p_A^* - p_B^*) x_A} =$$

$$\frac{x_A \cdot \left(\dfrac{p_A^*}{p_B^*}\right)}{1 + \left(\left(\dfrac{p_A^*}{p_B^*}\right) - 1\right) x_A} \tag{7.46}$$

Figure 7.15 shows that the equilibrium y_A becomes much greater than x_A as the p_A^*/p_B^* ratio increases. This is the underlying reason for the usefulness of distillation to concentrate readily vaporizing substances, i.e. substances with a high p_A^*. Based on such data, isothermal p-molar ratio data as shown in Fig. 7.16a can be obtained. The upper straight line (in blue) shows the total pressure as a function of the molar ratio (x_B) according to Raoult's law. The lower red line indicates the molar ratio of B in the gas phase (y_B). Let us follow the path A–D in Fig. 7.16a. At point A, only a single liquid solution phase is present. When the pressure is reduced, reaching point B, an additional phase is formed, gas richer at B (point Bg), which coexists with the liquid phase at point B. A further reduction in pressure to point C causes a transformation of the liquid to a gas; in this gas is richer in B than the liquid phase. The transformation of the liquid phase to gas is completed at the red curved line, and at point D only a single gas phase exists.

Figure 7.16b shows an isobaric T – molar ratio (x_B) phase diagram, with the liquid phase in the lower left-hand corner and the gas in the upper right-hand corner. If the liquid is heated following the path starting at point A upwards, point B is reached, and if the temperature is kept constant, a gas with a higher B-concentration is created (point C). This gas is condensed by cooling the gas (point D). By evaporating the

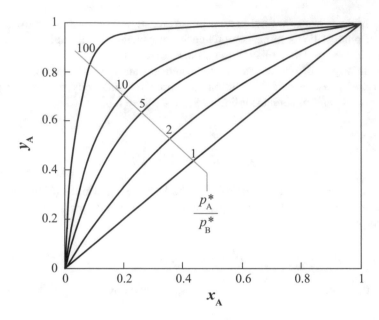

Fig. 7.15 Molar ratio of A in the gas phase (y_A) at equilibrium as a function of the molar ratio of A in the binary liquid (x_A). The data are isothermal and based on Eq. (7.46) with different $p_A{}^*/p_B{}^*$-values

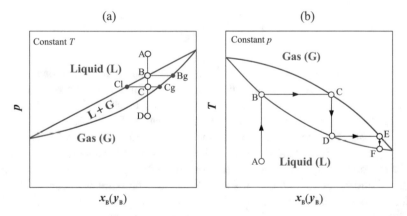

Fig. 7.16 Isothermal (**a**) and isobaric (**b**) phase diagrams of a binary system focusing on the liquid and gas (vapour) phases

liquid at constant temperature, a gas is formed with a high B-concentration (point E), which can be condensed (point F). This two-stage process (B → C and D → E) yield a liquid with a high B content.

In summary, a *complete phase diagram* including the three very different states of matter – solid, liquid and gas – is shown in Fig. 7.17. In the bottom part, two

Fig. 7.17 Generic isobaric phase diagram for a binary system including solid, liquid and gas phases

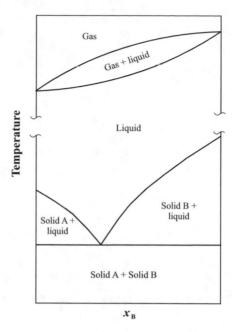

crystalline phases are present. *Crystallization* is a very demanding process, and it is a common feature that the other substance is prohibited from cocrystallizing. Some substances have an irregular structure and consequently they cannot crystallize. Instead they form a *glass*, which has a frozen-in amorphous structure. Melting point depression by adding small amounts of the other substance to the liquid phase is the "mother" of the eutectic point. The phase diagram presented is a simple one. It is assumed that the compounds are fully miscible. Many combinations of two substances show several phases (UCST or LCST behaviour). *Evaporation* is possible in the case of low molar mass species. Some compounds (polymers are a prominent example) degrade before they evaporate. Four of the regions contain two phases, whereas two domains have only one phase.

Chapter 8
Chemical Equilibrium

A central part of chemical thermodynamics is concerned with chemical reactions and *chemical equilibrium*. This is often treated in a separate course without the strictness typical of thermodynamics. In this final chapter, we shall deal with this topic using *Gibbs free energy* and related state functions to describe chemical equilibrium.

Let us start with a very simple reaction: A (g) $\leftrightarrows$ B (g). The changes in the number of moles of A (n_A) and B (n_B) with the extent of the forward reaction, which denoted the *reaction coordinate* (ξ), are given by:

$$dn_A = -d\xi; dn_B = d\xi \qquad (8.1)$$

The differential change in Gibbs free energy at constant T and constant p is given by:

$$dG = -\mu_A d\xi + \mu_B d\xi = (\mu_B - \mu_A)d\xi \qquad (8.2)$$

which means that:

$$\left(\frac{\partial G}{\partial \xi}\right)_{p,T} = \mu_B - \mu_A \qquad (8.3)$$

This derivative, which is also a difference, is abbreviated ΔG_r:

$$\Delta G_r = \left(\frac{\partial G}{\partial \xi}\right)_{p,T} = \mu_B - \mu_A \qquad (8.4)$$

and, according to Chap. 7 (Eq. (7.8)), it can be expressed as:

Fig. 8.1 Gibbs free energy
(G) at constant T and
constant p as a function of
the percentage of the extent
of the forward reaction (ξ).
Domains where the forward
reaction is dominant
($\Delta G_r < 0$) and where the
backward reaction is
dominant ($\Delta G_r > 0$). At
equilibrium, $\Delta G_r = 0$

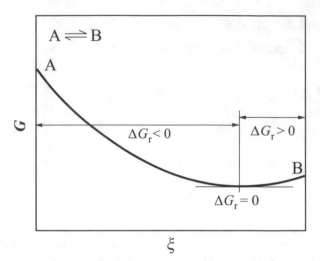

$$\Delta G_r = \mu_B - \mu_A = \left(\mu_B^* - \mu_A^*\right) + RT \ln\left(\frac{a_B}{a_A}\right) \qquad (8.5)$$

where the *activity ratio*, a_B/a_A, is denoted the *reaction quotient* (Q). Figure 8.1
shows a generic $G–\xi$ diagram. The curvature, which originates from the logarithmic
term in Eq. (8.5), is the reason for the presence of a minimum point, chemical
equilibrium in the $G–\xi$ diagram.

On the left-hand side of the minimum, $\Delta G_r < 0$, which indicates that the forward
reaction (A $\rightarrow$ B) dominates. This continues until the point where $\Delta G_r = 0$. If, on the
other hand, the system is on the right-hand side of the minimum, $\Delta G_r < 0$ and the
backward reaction (B $\rightarrow$ A) is dominant. Again, the system is striving to reach the
minimum G-value ($\Delta G_r = 0$). The *condition for equilibrium* can be expressed as:

$$\Delta G_r = 0 \qquad (8.6)$$

The difference between the chemical potentials of the pristine compounds A and
B can be written:

$$\mu_B^* - \mu_A^* = \Delta G_{f,B}^* - \Delta G_{f,A}^* = \Delta G_r^* \qquad (8.7)$$

where $\Delta G_{f,A}^*$ and $\Delta G_{f,B}^*$ are the free energy changes associated with the formation
of compounds A and B, respectively, and the difference is denoted ΔG_r^*.

At equilibrium, i.e. when $\Delta G_r = 0$:

Fig. 8.2 Free energy at constant T and constant p as a function of the extent of forward reaction (ξ). The free energies of formation of compounds A ($\Delta G_{f,A}{}^*$) and B ($\Delta G_{f,B}{}^*$) are shown. The curvature is obtained from the mixing of the compounds (ideal solution behaviour)

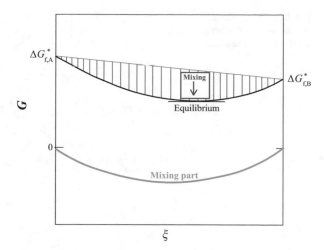

$$\Delta G_r^* + RT \ln Q(\mathrm{eq}) = 0 \Rightarrow K = e^{-\frac{\Delta G_r^*}{RT}} \tag{8.8}$$

where K is the *equilibrium constant*:

$$K = \left(\frac{a_B}{a_A}\right)_{eq} \tag{8.9}$$

The thermodynamic equilibrium constant is determined by a single structural parameter (ΔG_r^*); a high, positive value means that K is very small, close to zero. A large negative ΔG_r^* value means, on the other hand, that K is very high (Eq. (8.10)). The thermodynamic equilibrium constant has *no unit*, because activities are unit-free.

$$\Delta G_r^* = 0 \Rightarrow K = 1$$
$$\Delta G_r^* < 0 \Rightarrow K > 1 \tag{8.10}$$
$$\Delta G_r^* > 0 \Rightarrow K < 1$$

Figure 8.2 shows schematically the two factors determine the value of K: (i) the relationship between $\Delta G_{f,A}{}^*$ (reactant) and $\Delta G_{f,B}{}^*$ (product); if ΔG_f^* (reactants) is greater than ΔG_f^* (products), then $K > 1$ and (ii) the thermodynamics of the mixing of the two components, i.e. whether or not they are fully miscible. The calculation for more complex reactions requires the use of *stoichiometric numbers*, whose definition is indicated by the example shown in Fig. 8.3.

For more complex reactions, the following expressions are used:

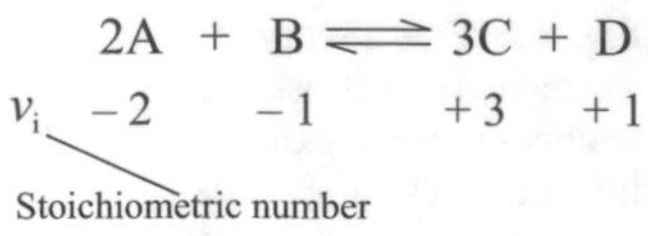

Fig. 8.3 Illustration of
stoichiometric number (v_i)
for a schematic reaction

$$2A \ + \ B \rightleftharpoons 3C \ + \ D$$

$$v_i \quad -2 \qquad -1 \qquad +3 \quad +1$$

Stoichiometric number

$$\Delta G_r = \Delta G_r^* + RT \ln Q = \sum_i v_i \Delta G_{f,i}^* + RT \ln Q \tag{8.11}$$

$$dG = \sum_i \mu_i dn_i = \left(\sum_i \mu_i v_i\right) d\xi \tag{8.12}$$

$$\Delta G_r = \left(\frac{\partial G}{\partial \xi}\right)_{p,T} = \sum_i \mu_i v_i = \sum_i \mu_i^* v_i + RT \sum_i v_i \ln a_i \tag{8.13}$$

$$\Delta G_r^* + RT \sum_i \ln a_i^{v_i} = 0 \Rightarrow \sum_i \ln a_i^{v_i} = -\frac{\Delta G_r^*}{RT}$$

$$\prod_i a_i^{v_i} = e^{-\frac{\Delta G_r^*}{RT}} \Rightarrow K = e^{-\frac{\Delta G_r^*}{RT}} \tag{8.14}$$

The thermodynamic equilibrium constant (K) can thus be given the following general form:

$$K = \prod_i a_i^{v_i} \tag{8.15}$$

The temperature dependence of K is expressed by the *van't Hoffs equation* on the basis of Eqs. (8.14) and (8.16).

$$\Delta G_r^* = \Delta H_r^* - T\Delta S_r^* \tag{8.16}$$

where ΔH_r^* and ΔS_r^* are assumed to be constant, which is reasonable for a moderately wide temperature range. The temperature-dependent equilibrium constant is given by:

$$K = e^{-\frac{\Delta H_r^*}{RT}} \cdot e^{\frac{\Delta S_r^*}{R}} \tag{8.17}$$

and its logarithm is:

$$\ln K = -\frac{\Delta H_r^*}{RT} + \frac{\Delta S_r^*}{R} \tag{8.18}$$

The temperature derivative of $\ln K$ is one of the formulations of the van't Hoff equation (named after the Dutch chemist Jacobus Henricus van't Hoff):

Fig. 8.4 Temperature dependence of the equilibrium constant (K) for an exothermal reaction (red line) and an endothermal reaction (blue line). This sketch is for a reaction with constant ΔH_r^* and constant ΔS_r^*

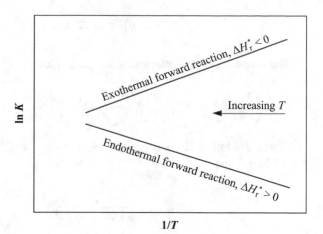

$$\frac{d \ln K}{dT} = \frac{\Delta H_r^*}{RT^2} \tag{8.19}$$

Another formulation of the van't Hoff equation is:

$$\ln K = C - \frac{\Delta H_r^*}{RT} \Rightarrow \frac{d \ln K}{d(1/T)} = -\frac{\Delta H_r^*}{R} \tag{8.20}$$

Figure 8.4 shows the influence of temperature on K for two classes of reaction: *exothermal* and *endothermal* forward reactions. The exothermal processes show a decrease in K with increasing temperature, whereas the endothermal processes show the opposite behaviour. *Athermal* reactions show a temperature-independent K.

The original work on $K = f(T)$ did not include the fact that both ΔH_r^* and ΔS_r^* are *temperature-dependent*, which has a significant impact on the results obtained when wide temperature ranges are involved. The temperature dependence of the reaction enthalpy is described by the *Kirchhoff law* (Chap. 1, Eq. (1.13)):

$$\Delta H_r^* = \Delta H_{r,0}^* + \Delta C_p T \tag{8.21}$$

where:

$$\Delta C_p = C_{p,p} - C_{p,r} \tag{8.22}$$

where $C_{p,p}$ and $C_{p,r}$ are the heat capacities at constant pressure of the products and reactants, respectively. Note that these require the use of the stoichiometric numbers: $\Delta C_p = \sum \nu_i \cdot C_{p,i}$. The following expression (based on Eq. (8.18)) is a convenient starting point for the derivation:

$$-R \ln K = \frac{\Delta H_r^*}{T} - \Delta S_r^* \tag{8.23}$$

The temperature derivatives of this equation are:

$$-R\left(\frac{\partial \ln K}{\partial T}\right) = \Delta H_r^* \cdot \left(\frac{\partial(1/T)}{\partial T}\right) + \frac{1}{T} \cdot \left(\frac{\partial(\Delta H_r^*)}{\partial T}\right) - \left(\frac{\partial(\Delta S_r^*)}{\partial T}\right) \tag{8.24}$$

The temperature dependence of the entropy is $\partial S/\partial T = C_p/T$ (Chap. 2, Eq. (2.12)). Thus,

$$\frac{\partial(\Delta S_r^*)}{\partial T} = \frac{\Delta C_p}{T} \tag{8.25}$$

which can be inserted in Eq. (8.24) to give:

$$-R\left(\frac{\partial \ln K}{\partial T}\right) = \frac{-\Delta H_r^*}{T^2} + \frac{\Delta C_p}{T} - \frac{\Delta C_p}{T} = \frac{-\Delta H_r^*}{T^2} \tag{8.26}$$

and the final expression becomes:

$$\left(\frac{\partial \ln K}{\partial T}\right) = \frac{\Delta H_r^*(T)}{RT^2} \quad ; \quad \left(\frac{\partial \ln K}{\partial(1/T)}\right) = \frac{-\Delta H_r^*(T)}{R} \tag{8.27}$$

Equation (8.27) appears to be very similar to the original van't Hoff equation (Eq. 8.20), but in Eq. (8.27), ΔH_r^* is a function of T:

$$\Delta H_r^*(T) = A + \Delta C_p T \tag{8.28}$$

where A is a constant. Figure 8.5 shows the effect of a temperature-dependent ΔH_r^* on the $\ln K - 1/T$ plot.

The equilibrium constant has no explicit *dependence on pressure*, but if a gas mixture with an established chemical equilibrium between compounds A and B is compressed according to Fig. 8.6, the reaction conditions are changed and the system reacts, despite the fact that K is independent of pressure. A reduction in volume from V to $V/2$ at constant temperature means that all partial pressures are doubled ($pV = $ constant):

$$p_{new} = 2p_{eq} \tag{8.29}$$

The equilibrium constant K of the reaction A $\leftrightarrows$ 2B is:

Fig. 8.5 Temperature
dependence of the
equilibrium constant (K) for
an endothermal reaction
with constant ΔH_r^* and
constant ΔS_r^* (black straight
line) and for a reaction with
temperature-dependent
ΔH_r^* and ΔS_r^* (red curved
line)

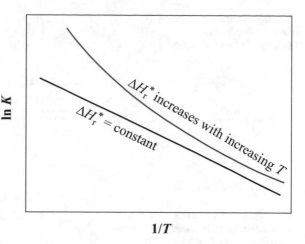

Fig. 8.6 (a) Equilibrium
between the gases A and B
according to the displayed
reaction established at a
specific volume (V); (b) the
gas mixture is compressed
to half the original volume
($V \rightarrow V/2$); the effect of this
action is discussed in
the text

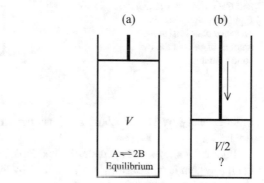

$$K = \frac{p_{B,eq}^2}{p_{A,eq} \cdot p^*} \qquad (8.30)$$

where p^* is a standard pressure (constant). The compression causes, however, a change in the reaction quotient (Q) from K to:

$$Q = \frac{4 \cdot p_{B,eq}^2}{2 \cdot p_{A,eq} \cdot p^*} = 2K \qquad (8.31)$$

The effect of the compression is thus that the backward reaction (2B→A) will dominate until Q is equal to K. This reaction, in which two molecules associate into one molecule, causes a reduction in the number of molecules in the gas system; this in turn reduces the pressure, since p is proportional to n at constant T and V. This is one example of the *Le Chatelier-Braun principle*: to an action, in this case an increase in pressure, the system response is a re-action, i.e. a counteraction; in this particular case, a decrease in the pressure. This important principle was formulated

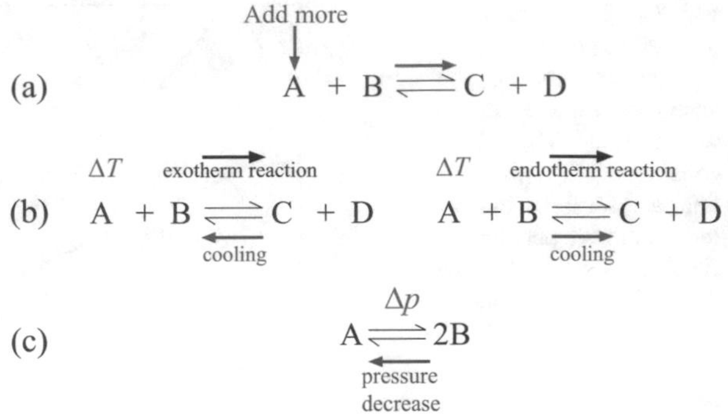

Fig. 8.7 Illustration of the Le Chatelier-Braun principle in three examples

Add more

(a) $A + B \rightleftharpoons C + D$

ΔT exotherm reaction ΔT endotherm reaction

(b) $A + B \rightleftharpoons C + D$ $A + B \rightleftharpoons C + D$

cooling   cooling

Δp

(c) $A \rightleftharpoons 2B$

pressure
decrease

Fig. 8.8 A heterogeneous equilibrium including solid-state compounds (CaCO$_3$ and CaO) and a gas (CO$_2$). Equilibrium requires the presence of at least some amount of each of the two solid compounds

$$CaCO_3 \text{ (s)} \rightleftharpoons CaO \text{ (s)} + CO_2 \text{ (g)}$$

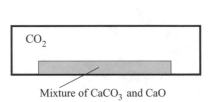

Mixture of CaCO$_3$ and CaO

independently by Henry Louis Le Chatelier (in 1885) and Karl Ferdinand Braun (in 1886).

Figure 8.7 shows a few illustrative examples of the Le Chatelier-Braun principle: (a) If a system in equilibrium is "disturbed" by the feeding of more substance A, then the forward reaction (marked with a red arrow) will dominate until equilibrium is attained. Hence, the primary reaction of the system is to reduce the concentration of substance A, a counteraction of the first action. (b) A disturbance by a temperature increase causes a reaction from the system by cooling. According to the van't Hoff equation, an exothermal reaction reacts by decreasing K, which means that the backward reaction dominates. The backward reaction is endothermal, and this leads to a cooling of the system. If the forward reaction is endothermal, then K increases with increasing temperature and the dominant forward reaction cools the system. (c) The effect of compression of a reaction gas mixture has already been discussed.

Chemical equilibria are also established between substances in different aggregation states. One such example is shown in Fig. 8.8, where two solid phase compounds (CaCO$_3$ (s) and CaO (s)) and a gas, CO$_2$, are involved. The equilibrium constant is, according to the principles, compactly described by Eq. (8.15):

$$K = \frac{a_{CO_2}}{a_{CaCO_3 \text{(s)}} a_{CaO \text{(s)}}} = a_{CO_2} = \frac{p_{CO_2}}{p_{CO_2}^*} \qquad (8.32)$$

The activities of the pure solid phases are, by definition, unity, which means that the expression contains only one activity, the activity of the carbon dioxide, and that the presence of 10 mg or 1000 kg of solid phase at equilibrium has no impact on the partial pressure of carbon dioxide.

Chapter 9
Thermodynamics Problems

Gases

1. Derive an expression for the volumetric expansion coefficient (α) for an ideal gas; α is given by: ($\alpha = (1/V)(\partial V/\partial T)_p$.
2. Derive an expression for the isothermal compressibility (κ_T) for an ideal gas; κ_T is given by: $\kappa_T = (-1/V)(\partial V/\partial p)_T$.

First Law

3. For an ideal gas, the following relation holds: $(\partial U/\partial V)_T - 0$. Derive the following relationships from this expression: (a) $(\partial U/\partial p)_T = 0$; (b) $(\partial H/\partial V)_T = 0$; (c) $(\partial H/\partial p)_T = 0$; (d) $(\partial T/\partial p)_H = 0$.
4. Show that $C_{p,m} - C_{v,m} = R$ for an ideal gas.
5. The heat capacities at constant pressure for $CHCl_3$ are 118 J $(K\ mol)^{-1}$ (liquid) and 71 $(K\ mol)^{-1}$ (gas). The evaporation enthalpy ($\Delta H_{vap,m}$) at 60 °C is 29.20 kJ mol^{-1}. Calculate $\Delta H_{vap,m}$ at 20 °C and 100 °C.
6. Calculate ΔH^0 and ΔU^0 for hydrogenation of acetylene to ethylene at 25 °C and 1 bar based on ΔH^0 data of the following reactions:

$$H_2(g) + \tfrac{1}{2}O_2(g) \rightarrow H_2O(l)\ \Delta H^0_1 = -286\ kJ\ mol^{-1}$$
$$C_2H_2(g) + 2\tfrac{1}{2}O_2(g) \rightarrow H_2O(l) + 2CO_2(g)\ \Delta H^0_2 = -1300\ kJ\ mol^{-1}$$
$$C_2H_4(g) + 3O_2(g) \rightarrow 2H_2O(l) + 2CO_2(g)\ \Delta H^0_3 = -1410\ kJ\ mol^{-1}$$

First and Second Laws Combined

7. 50 g oxygen gas (298 K and 1 bar) expands reversibly and adiabatically to the double volume. Calculate the changes in internal energy (ΔU) and entropy (ΔS). Calculate also the entropy production (ΔS_i). Assume that the gas follows the ideal gas law.

8. Make a calculation for the same gas (298 K and 1 bar). In this case, the gas expands into an evacuated container (Joule expansion). The volume of the gas is doubled by the expansion. Calculate ΔU, ΔS and ΔS_i.

9. Water drops at $-15\ ^{\circ}C$ can exist for some time in a pure state. However, ultimately when touching ground, they freeze immediately making walking and driving a car a very dangerous activity. Show based on the second law that crystallization of liquid water at $-10\ ^{\circ}C$ is a spontaneous process. The following data are needed to solve the problem: $C_{p,m}$ (water) $= 75.3\ J\ (K\ mol)^{-1}$, $C_{p,m}$ (ice) $= 37.7\ J\ (K\ mol)^{-1}$, $\Delta H_{m,m} = 6024\ J\ mol^{-1}$.

10. Show that the change in entropy (ΔS) of an ideal gas in V–T space can be described by $\Delta S = C_v \ln (T_2/T_1) + nR \ln (V_2/V_1)$.

11. Use the derived relationship together with other suitable equations to determine ΔS for an ideal gas subjected to the following changes: (i) reversible isothermal expansion; (ii) reversible isochoric heating; (iii) isobaric change of state; (iv) reversible adiabatic expansion; and (v) free adiabatic expansion (Joule expansion).

Gibbs and Helmholtz Free Energies and the Maxwell Relations

12. Calculate the change in Gibbs free energy of melting of ice at normal pressure at the following temperatures: $-10\ ^{\circ}C$, $0\ ^{\circ}C$, $10\ ^{\circ}C$ and $30\ ^{\circ}C$. The enthalpy of melting is $6024\ J\ mol^{-1}$, and the entropy of melting is $22.05\ J\ (K\ mol)^{-1}$.

13. Derive the following relationships for the Gibbs free energy: $(\partial G/\partial T)_{p,n} = -S$, $(\partial G/\partial p)_{T,n} = V$ and $(\partial G/\partial n)_{T,p} = \mu$.

14. Show that the internal pressure (π_T) of an ideal gas is zero by applying one of the Maxwell relations on the combined first and second laws.

Chemical Potential and Phase Equilibria

15. At the melting point (1 bar), $-38.9\ ^{\circ}C$, the densities of, respectively, liquid and solid mercury are $13,690\ kg\ m^{-3}$ and $14,193\ kg\ m^{-3}$. The melting enthalpy of mercury at this temperature is $9750\ J\ (kg)^{-1}$. Determine the melting point of mercury at 3549 bar. The experimentally determined melting point of mercury at

3540 bar is $-19.9\,°C$. Please, comment about the difference and suggest an explanation to the discrepancy.

16. In order to make medical instruments sterile, they have to be treated at $120\,°C$ to eliminate the spores of certain bacteria. Which pressure is required to reach a boiling point of water at $120\,°C$. The enthalpy of evaporation of water is $40,600\ \mathrm{J\ mol^{-1}}$.

17. The vapour pressure above solid and liquid HCN is given by:

$$\log p = 9.3390 - \frac{1864.8}{T} \quad \text{(solid Hg)} \tag{9.1}$$

$$\log p = 7.7446 - \frac{1453.1}{T} \quad \text{(liquid Hg)} \tag{9.2}$$

Determine the temperature and pressure at which three phases (solid, liquid and gas; i.e. the triple point) coexists.

Colligative Properties

18. The partial pressure of water (p_{H2O}; given in mm Hg) shows the following dependence of the content of a second component B at $40\,°C$. The solution behaves ideally (according to Raoult's law) up to a certain molar ratio of B (x_B). At which x_B-value is this? A second question: which is the osmotic pressure at $x_B = 0.2$?

x_{H2O}	1	0.98	0.96	0.94	0.90	0.88	0.86	0.82	0.80	0.78
p_{H2O}	55.3	54.2	53.1	52.1	50.7	50.2	49.5	49.4	48.7	47.1

19. A protein weighing 200 mg is dissolved in 10 mL water at $25\,°C$. The osmotic pressure is measured to 4000 Pa. Calculate the molar mass of the protein assuming that the solution is ideal.

20. Determine the osmotic pressure in a water solution of KCl at $25\,°C$ from the following partial water pressure data: 3200 Pa (above pure water) and 2600 Pa (above the solution).

21. Calculate the melting point of water solutions of NaCl with the following mass percentages of NaCl: 1, 2, 5 and 10 wt.%. The molar melting enthalpy of water is $6024\ \mathrm{J\ mol^{-1}}$. Assume that the solutions are ideal.

Chemical Equilibrium

22. The following gas phase reaction – A $\leftrightarrows$ 3B – is allowed to reach chemical equilibrium. After this, the system is allowed to expand from the initial volume V_0 to $5V_0$. What will happen after this action?

23. Determine the equilibrium constant for the following reaction at 298 K:

$$\tfrac{1}{2} H_2(g) + \tfrac{1}{2}Cl_2(g) \leftrightarrows HCl(g)$$

The ΔG_f^0 for the formation of HCl (g) is $-95,300$ J mol^{-1}.

24. The dissociation of methane ($CH_4(g)$) to hydrogen ($H_2(g)$) and graphite ($C(s)$) is the topic of this problem. (i) Calculate the equilibrium constant (K) at 298 K based on the following data: $\Delta H_f^\circ = -74,850$ J mol^{-1} and $\Delta S_f^\circ = -80.67$ J K^{-1} mol^{-1}. (ii) Assume that ΔH_f° is independent of temperature and calculate K at 323 K. (iii) Calculate the degree of dissociation of methane (α) at 298 K and the total pressure of 0.01 bar.

Phase Diagrams

25. Mark the phases present in each of the areas of the isobaric binary phase diagram (consisting of substances A and B) below. Mark also the eutectic point and the compositions of the phases present in each of three marked dots. Calculate the relative amounts of each the six phases (Fig. 9.1).

Fig. 9.1 Phase diagram consisting of two substances (A and B) at constant pressure

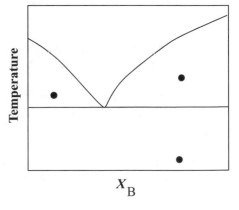

Chapter 10
Solutions to Problems

1. The thermal expansion coefficient is given by:

$$\alpha = \frac{1}{V}\left(\frac{\partial V}{\partial T}\right)_p \tag{10.1}$$

An ideal gas follows the ideal gas law:

$$pV = nRT \Rightarrow V = \left(\frac{nR}{p}\right) \cdot T$$

$$\left(\frac{\partial V}{\partial T}\right)_p = \frac{nR}{p} \tag{10.2}$$

which is inserted in Eq. (10.1) gives:

$$\alpha = \frac{nR}{pV} = \frac{1}{T} \Leftrightarrow T \to \infty \Rightarrow \alpha = 0 \tag{10.3}$$

2. The isothermal compressibility is given by:

$$\kappa_T = -\left(\frac{1}{V}\right) \cdot \left(\frac{\partial V}{\partial p}\right)_T \tag{10.4}$$

The ideal gas law is reformulated into:

$$pV = nRT \Rightarrow V = (nRT) \cdot \frac{1}{p}$$

$$\left(\frac{\partial V}{\partial p}\right)_T = -\frac{nRT}{p^2} \tag{10.5}$$

which by insertion into Eq. (10.4) gives:

$$\kappa_T = \frac{nRT}{p^2 V} = \frac{nRT}{pV} \cdot \frac{1}{p} = \frac{1}{p} \Leftrightarrow p \to \infty \Rightarrow \kappa_T \to 0 \tag{10.6}$$

3. (a) The internal energy of an ideal gas is solely a function of T, which immediately implies that the derivatives of U with respect to either V or p while keeping T constant must be zero. More formally:

$$dU = \left(\frac{\partial U}{\partial T}\right)_V dT + \left(\frac{\partial U}{\partial V}\right)_T dV$$

$$\left(\frac{\partial U}{\partial V}\right)_T = 0 \ (\text{ideal gas}) \Rightarrow dU = \left(\frac{\partial U}{\partial T}\right)_V dT \tag{10.7}$$

which provides the input yielding the final expression:

$$\therefore U = f(T) \Rightarrow \left(\frac{\partial U}{\partial p}\right)_T = 0 \tag{10.8}$$

(b) Starting with the definition of H and then by differentiating this expression with respect to V while keeping T constant, the following expressions are obtained:

$$H = U + pV$$

$$\left(\frac{\partial H}{\partial V}\right)_T = \left(\frac{\partial U}{\partial V}\right)_T + \left(\frac{\partial (pV)}{\partial V}\right)_T \tag{10.9}$$

The last term in this expression is part of the ideal gas law:

$$pV = nRT \tag{10.10}$$

which is inserted into Eq. (10.9):

$$\left(\frac{\partial H}{\partial V}\right)_T = \left(\frac{\partial U}{\partial V}\right)_T + \left(\frac{\partial (nRT)}{\partial V}\right)_T = 0 + 0 = 0 \tag{10.11}$$

(c) By differentiating the defining expression of H with respect to p while keeping T constant, the following expression is obtained:

$$\left(\frac{\partial H}{\partial p}\right)_T = \left(\frac{\partial U}{\partial p}\right)_T + \left(\frac{\partial (pV)}{\partial p}\right)_T = \left(\frac{\partial U}{\partial p}\right)_T + \left(\frac{\partial (nRT)}{\partial p}\right)_T$$

$$\left(\frac{\partial H}{\partial p}\right)_T = 0 + 0 = 0 \tag{10.12}$$

(d) Isoenthalpic conditions indicate that the total differential of H is zero:

$$dH = dU + d(nRT) = 0 \ (H = \text{constant})$$

$$\left(\frac{\partial U}{\partial p}\right)_H + nR\left(\frac{\partial T}{\partial p}\right)_H = 0 \Rightarrow 0 + nR\left(\frac{\partial T}{\partial p}\right)_H = 0 \tag{10.13}$$

$$\therefore \left(\frac{\partial T}{\partial p}\right)_H = 0$$

4. The derivation starts with the definition of H; this expression is differentiated with respect to T maintaining p constant:

$$H = U + pV$$

$$\left(\frac{\partial H}{\partial T}\right)_p = \left(\frac{\partial U}{\partial T}\right)_p + \left(\frac{\partial (pV)}{\partial T}\right)_p = \left(\frac{\partial U}{\partial T}\right)_p + \left(\frac{\partial (nRT)}{\partial T}\right)_p \tag{10.14}$$

For an ideal gas, U is only a function of T, which can be used to show that C_v has an alternative definition for an ideal gas:

$$dU = C_v dT$$

$$\left(\frac{\partial U}{\partial T}\right)_p = C_v \cdot \left(\frac{\partial T}{\partial T}\right)_p = C_v \tag{10.15}$$

which is inserted in Eq. (10.14):

$$C_p = C_v + nR$$

$$\frac{C_p}{n} = \frac{C_v}{n} + R \Rightarrow C_{p,m} - C_{v,m} = R \tag{10.16}$$

According to statistical thermodynamics, $C_{v,m}$ is $3R/2$ for a monoatomic gas (i.e. the noble gases) and $5R/2$ for diatomic gases.

5. The Kirchhoff law is used with the approximation that the heat capacities are constant, independent of temperature:

$$\Delta H(20\,^\circ\text{C}) = \Delta H(60\,^\circ\text{C}) + \int_{60}^{20} \left(C_{p,p} - C_{p,r}\right) dT \tag{10.17}$$

$$\Delta H(20\,^\circ\text{C}) = 29200 + (71 - 118) \cdot (20 - 60) = 31,680\ \text{J/mol}$$

Note that integral is simplified to $\Delta C_p \cdot \Delta T$.

6. The reaction is C_2H_2 (g) + H_2(g) $\rightarrow$ C_2H_4(g), which is equal to the sum of reaction 1 (oxidation of hydrogen), reaction 2 (oxidation of acetylene) and the backward reaction of reaction 3 (oxidation of ethylene). According to Hess law, $\Delta H^0 = \Delta H^0{}_1 + \Delta H^0{}_2 - \Delta H^0{}_3 = -176$ kJ mol^{-1}. The change in internal energy under standards conditions (25 °C and 1 bar) is obtained from $\Delta H^0 = \Delta U^0 + \Delta(pV) = \Delta U^0 + \Delta(nRT) = \Delta U^0 + RT(\Delta n) \Rightarrow \Delta U^0 = \Delta H^0 - RT$ $(\Delta n) = -176{,}000 - 8.314 \cdot 298 \cdot (-1) = -173.5$ kJ mol^{-1}.

7. First task is to calculate the number of moles of O_2: $n = 50/32 = 1.56$ mol.

   A reversible, adiabatic expansion causes a temperature decrease:

$$T_2 = T_1 \cdot \left(\frac{V_2}{V_1}\right)^{-\frac{R}{C_{v,m}}} \tag{10.18}$$

   where $C_{v,m}$ is the molar heat capacity. The following data are inserted to calculate the final temperature:

$$T_2 = 298 \cdot 2^{-8.34/20.7} = 225.6\ \text{K}$$

$$\Delta U = C_{v,m} \cdot n \cdot (T_2 - T_1) = 20.7 \cdot 1.56 \cdot (225.6 - 298) = -2338\ \text{J}$$

   The process is adiabatic and reversible: $\Delta S = 0$ and $\Delta S_i = 0$.

8. Joule expansion: $\Delta U = w + q = 0$ because $p_{ex} = 0$ and the heat transfer is negligible, $q = 0$. The fact that for an ideal gas $dU = C_v\, dT$ implies that T is constant. ΔS is calculated from isothermal, reversible expansion from V to $2V$.

$$\Delta S_{\text{system}} = nR \ln (V_2/V_1) = 1.56 \cdot 8.314 \cdot \ln 2 = 8.99\ \text{J K}^{-1}.$$

   The entropy production (ΔS_i) is the sum of the entropy changes in the system (8.99 J K^{-1}) and in the surroundings (0 J K^{-1}): $\Delta S_i = 8.99$ J K^{-1}. The positive entropy production indicates that this process is spontaneous.

9. The change in entropy associated with crystallization at -15 °C is obtained by finding a reversible path: first heating water from -15 °C to 0 °C, transforming water to ice at 0 °C and finally cooling ice from 0 °C to -15 °C. The calculation is based on 1 mole of matter:

$$\Delta S_{syst} = 75.3 \cdot \ln\ (273.2/258.2) + (-6024)/273.2 + 37.7 \cdot \ln(258.2/273.2)$$
$$= -19.93 \text{ J K}^{-1}$$

The change in entropy of the surroundings requires calculation of the enthalpy of melting using the Kirchhoff law: $\Delta H_{m,m}$ (-15 °C) = 6024–15· (75.3–37.7) = 5460 J mol^{-1}; this heat is transferred to the surrounding in a reversible fashion at –15 °C: ΔS_{surr} = 5460/258.2 = +21.15 J K^{-1}.
The entropy production $\Delta S_i = \Delta S_{syst} + \Delta S_{surr} = -19.93 + 21.15 = +1.22$ J K^{-1}.
The fact that ΔS_i is greater than zero shows that the process is spontaneous.

10. The state change is divided into two reversible movements in V–T space. First an isothermal change in gas volume from V_1 to V_2:

$$dU = dq + dw = 0 \Rightarrow dq = -dw = pdV$$
$$dq = nRT \cdot \frac{dV}{V} \Rightarrow q = nRT \ln\left(\frac{V_2}{V_1}\right) \tag{10.19}$$
$$\Delta S_1 = \frac{q}{T} = nR \ln\left(\frac{V_2}{V_1}\right)$$

The second change is at constant volume (isochoric) involving a temperature change from T_1 to T_2:

$$dq = C_v dT \ \wedge\ dS = \frac{dq}{T} = C_v \cdot \frac{dT}{T}$$
$$\Delta S_2 = C_v \ln\left(\frac{T_2}{T_1}\right) \tag{10.20}$$

The sum of the two reversible processes provides the final equation:

$$\Delta S = \Delta S_1 + \Delta S_2 = nR \ln\left(\frac{V_2}{V_1}\right) + C_v \ln\left(\frac{T_2}{T_1}\right) \tag{10.21}$$

This expression can also be derived using a different path starting with the combined first and second laws:

$$dU = TdS - pdV \ (n = \text{constant}) \Rightarrow dS = \frac{dU}{T} + \frac{p}{T}dV \tag{10.22}$$

and then considering that U is only a function of T for an ideal gas and, in addition, the last term in Eq. (10.22) can be expressed differently by applying the ideal gas law:

$$dU = C_v dT$$
$$pV = nRT \Rightarrow \frac{p}{T} = \frac{nR}{V} \tag{10.23}$$

These expressions are inserted into Eq. (10.22) to obtain the final expression:

$$dS = \frac{C_v dT}{T} + \frac{nR}{V} dV \Rightarrow \Delta S = C_v \ln\left(\frac{T_2}{T_1}\right) + nR \ln\left(\frac{V_2}{V_1}\right) \qquad (10.24)$$

11. The relationships below are based on Eq. (10.21):

(i) Reversible isothermal expansion, $T_1 = T_2$:

$$\Delta S = nR \ln\left(\frac{V_2}{V_1}\right) \qquad (10.25)$$

(ii) Reversible isochoric heating means that $V_1 = V_2$ and according to Eq. (10.21), the following expression holds:

$$\Delta S = C_v \ln\left(\frac{T_2}{T_1}\right) \qquad (10.26)$$

(iii) Isobaric change of state and according to the ideal gas law the following useful equation can be derived:

$$pV = nRT \Rightarrow \frac{V}{T} = \frac{nR}{p} = \text{constant} \Rightarrow \frac{V_1}{T_1} = \frac{V_2}{T_2}$$
$$\frac{V_2}{V_1} = \frac{T_2}{T_1} \qquad (10.27)$$

which is inserted into Eq. (10.21) to give:

$$\Delta S = nR \ln\left(\frac{T_2}{T_1}\right) + C_v \ln\left(\frac{T_2}{T_1}\right) = (C_v + nR) \ln\left(\frac{T_2}{T_1}\right) =$$
$$\Delta S = C_p \ln\left(\frac{T_2}{T_1}\right) \qquad (10.28)$$

Note that $C_p = C_v + nR$ according to Eq. (10.16).

(iv) The definition of dS provides the solution:

$$dq_{rev} = 0 \Rightarrow dS = \frac{dq_{rev}}{T} = 0 \Rightarrow \Delta S = 0 \qquad (10.29)$$

(v) Free expansion, an irreversible process, means that $\Delta U =$ constant and thus the expansion is isothermal for an ideal gas. Thus, Eq. (10.21) can be used under the condition of constant temperature, $T_1 = T_2$, which is the case of reversible isothermal expansion:

$$\Delta S = nR \ln \left(\frac{V_2}{V_1}\right) \tag{10.30}$$

12. The Gibbs free energy of melting is:

$$\Delta G_{m,m} = \Delta H_{m,m} - T \Delta S_{m,m} \tag{10.31}$$

$$\Delta G_{m,m} = 6024 - 22.05 \cdot T \tag{10.32}$$

The following $\Delta G_{m,m}$ values (in J mol^{-1}) were obtained: +220 (-10 °C), 0 (0 °C), -200 (10 °C) and -661 (30 °C). Melting is thus spontaneous, judged by the negative $\Delta G_{m,m}$-values at temperatures above 0 °C.

13. The starting point is the expression for dU, including both the first and second laws:

$$dU = TdS - pdV + \mu dn \tag{10.33}$$

The enthalpy is defined according to:

$$H = U + pV \tag{10.34}$$

which is differentiated to:

$$dH = dU + pdV + Vdp \tag{10.35}$$

and combined with Eq. (10.33):

$$dH = TdS - pdV + \mu dn + pdV + Vdp = Tds + Vdp + \mu dn \tag{10.36}$$

The Gibbs free energy is defined according to:

$$G = H - TS \tag{10.37}$$

which is differentiated as follows:

$$dG = Tds + Vdp + \mu dn - TdS - SdT = -SdT + Vdp + \mu dn \qquad (10.38)$$

The Gibbs free energy is thus:

$$G = f(T, p, n) \Rightarrow dG = \left(\frac{\partial G}{\partial T}\right)_{p,n} dT + \left(\frac{\partial G}{\partial p}\right)_{T,n} dp + \left(\frac{\partial G}{\partial n}\right)_{T,p} dn \qquad (10.39)$$

The partial derivatives can be identified by comparing Eqs. (10.38) and (10.39):

$$\left(\frac{\partial G}{\partial T}\right)_{p,n} = -S; \left(\frac{\partial G}{\partial p}\right)_{T,n} = V; \left(\frac{\partial G}{\partial n}\right)_{T,p} = \mu \qquad (10.40)$$

14. The first and seconds laws combined for a system with constant number of molecules can be expressed according to:

$$dU = TdS - pdV \text{ (constant } n) \qquad (10.41)$$

Derivation of this expression with respect to V while keeping T and n constant gives:

$$\left(\frac{\partial U}{\partial V}\right)_{T,n} = T\left(\frac{\partial S}{\partial V}\right)_{T,n} - p \qquad (10.42)$$

The differential expression for the Helmholtz free energy (A) at constant number of molecules is:

$$dA = -SdT - pdV \qquad (10.43)$$

By applying the Schwarz theorem (according to Maxwell), the following expression is obtained:

$$\left(\frac{\partial S}{\partial V}\right)_{T,n} = \left(\frac{\partial p}{\partial T}\right)_{V,n} \qquad (10.44)$$

The internal pressure (π_T) is given by (combining Eqs. (10.42) and (10.44)):

$$\pi_T = \left(\frac{\partial U}{\partial V}\right)_{T,n} = T\left(\frac{\partial p}{\partial T}\right)_{V,n} - p \qquad (10.45)$$

The ideal gas law is used to express p and the temperature derivative of p:

$$pV = nRT \Rightarrow p = \frac{nR}{V} \cdot T \wedge \left(\frac{\partial p}{\partial T}\right)_{V,n} = \frac{nR}{V} \qquad (10.46)$$

which are inserted into Eq. (10.45):

$$\pi_T = \left(\frac{\partial U}{\partial V}\right)_{T,n} = T \cdot \frac{nR}{V} - \frac{nR}{V} \cdot T = 0 \qquad (10.47)$$

15. The calculation starts with the Clapeyron equation:

$$\frac{dp}{dT} = \frac{\Delta S_{m,m}}{\Delta V_{m,m}} = \frac{\Delta H_{m,m}}{T \Delta V_{m,m}} \qquad (10.48)$$

which is integrated according to:

$$\int_{p^0}^{p} dp = \frac{\Delta H_{m,m}}{\Delta V_{m,m}} \cdot \int_{T^0}^{T} \frac{dT}{T} \Rightarrow \frac{\Delta V_{m,m}}{\Delta H_{m,m}} (p - p^0) = \ln\left(\frac{T}{T^0}\right) \Rightarrow$$

$$T = T^0 \exp\left(\frac{\Delta V_{m,m}}{\Delta H_{m,m}} (p - p^0)\right) \qquad (10.49)$$

The change in molar volume associated with melting is given by:

$$\Delta V_{m,m} = V_{m,l} - V_{m,s} = M\left(\frac{1}{\rho_l} - \frac{1}{\rho_s}\right) = 5.1775 \cdot 10^{-7} \ \text{m}^3 \ \text{mol}^{-1} \qquad (10.50)$$

and the molar enthalpy of melting is:

$$\Delta H_{m,m} = 9.75 \cdot 200 = 1950 \ \text{J mol}^{-1} \qquad (10.51)$$

which are inserted in Eq. (10.49) together with $p = 3.54 \cdot 10^8$ Pa, $p^0 = 10^5$ Pa and $T^0 = 234.4$ K $\Rightarrow T = 257.3$ K $= -15.8$ °C. The calculated temperature is thus higher than the experimentally determined melting point, -19.9 °C. The change in molar volume associated with melting ($\Delta V_{m,m}$) is expected to depend on pressure, which most probably account for the difference between the calculated and the experimental melting points.

16. The Clausius-Clapeyron equation (integrated form) can be used to determine the required pressure:

$$\ln(p/p^0) = \frac{\Delta H_{m,v}}{R}\left(\frac{1}{T^0} - \frac{1}{T}\right) = \frac{40600}{8.31} \cdot \left(\frac{1}{373} - \frac{1}{393}\right) = 0.6665 \qquad (10.52)$$

The searched pressure (p) is 1.95 atm $= 1.95 \cdot 10^5$ Pa.

Fig. 10.1 Partial pressure of water (in gas phase) as a function water molar ratio in solution

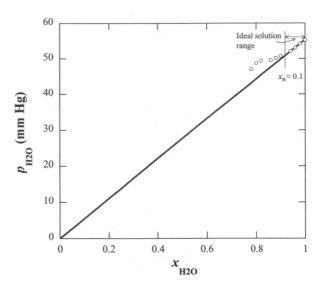

17. The triple point is obtained by setting Eq. (9.1) equal to Eq. (9.2):

    $9.3390–1864.8/T = 7.7446–1453.1/T \Rightarrow T = 258.8$ K, which after insertion in Eq. (9.1) gives $p = 137$ mm Hg $= 1.8 \cdot 10^4$ Pa.

18. The limits for ideal solution behaviour are shown in Fig. 10.1.

    The osmotic pressure is given by $\pi = (-RT \ln a_A)/V_A$; the molar volume of water is 18 g/mol/10^6 g/m^3 $= 1.8 \cdot 10^{-5}$ m^3 mol^{-1}. The activity at $x_B = 0.2$ is $p_{H2O}/p_{H2O}{}^* \approx 0.88 \Rightarrow \pi = -8.314 \cdot 313 \cdot \ln 0.88/(1.8 \cdot 10^{-5}) = 18.5$ MPa.

19. The osmotic pressure is given by $\pi = (RT/V_{m,H2O}) \cdot x_p$. The followingvalues are known: $\pi = 4000$ Pa, $R = 8.314$ J K^{-1} mol^{-1}, $T = 298$ K, $V_{m,H2O} = 18 \cdot 10^{-6}$ m^3 mol^{-1}. By inserting these values into the above equation, the molar ratio of the protein is obtained: $x_p = 2.906 \cdot 10^{-5}$. The molar mass of the protein is calculated as follows:

    $$2.906 \cdot 10^{-5} = 0.2/M_p/10/18 \Rightarrow M_p = 12,500 \text{ g mol}^{-1}.$$

20. The osmotic pressure is given by $\pi = -(RT/V_{m,H2O}) \cdot \ln a_{H2O}$. The activity of water in the solution is $a_{H2O} = p_{H2O}/p_{H2O}{}^* = 2600/3200 = 0.8125$. The osmotic pressure of the solution is thus:

    $$\pi = -8.314 \cdot 298 \cdot \ln (0.8125)/(18 \cdot 10^{-6}) = 25.6 \text{ MPa}.$$

21. The melting point depression for an ideal solution is given by:
    $\Delta T = (RT^2/\Delta H_{H2O,m,m}) \cdot x_{salt} = K_{\Delta T} \cdot x_{salt}$. The factor $K_{\Delta T}$ is

Fig. 10.2 Freezing point depression (ΔT) as a function of mass percentage of NaCl in water solution. The filled circles and the line are from calculation according to Eq. (7.41) assuming ideal solution behaviour. Experimental data (unfilled circles) are from Hall et al. (1988)

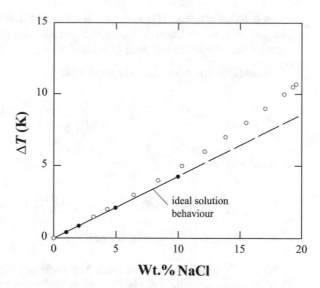

$8.31 \cdot 273^2/6024 \approx 103$ K. The mass percentages ($m\%$) have to be converted to molar ratios, $M(\text{H}_2\text{0}) \approx 18$ g mol^{-1}, $M(\text{NaCl}) \approx 58$ g mol^{-1}:

$$x_{\text{salt}} = m\%/M(\text{NaCl})/(m\%/M(\text{NaCl}) + (1 - m\%)/M(\text{H}_2\text{O})) \Rightarrow$$

$$x_{\text{salt}} = m\%/58/(m\%/58 + (1 - m\%)/18)$$

The following molar ratios are obtained: 0.003 (1 wt.%), 0.006 (2 wt.%), 0.016 (5 wt.%) and 0.033 (10 wt. %). The melting points of the different solutions become $-0.3\,^\circ$C (1 wt.%), $-0.6\,^\circ$C (2 wt.%), $-1.7\,^\circ$C (5 wt.%) and $-3.4\,^\circ$C (10 wt. %). A comparison with experimental data is shown in Fig. 10.2. The agreement between ideal solution behaviour and experimental data are relatively good for solution containing less than 5 wt. % NaCl. The eutectic point is at 23 wt.% NaCl showing a melting point depression of 21.2 K ($T_\text{m} = -21.2\,^\circ$C).

22. The equilibrium constant is given by:

$$K = \frac{a_\text{B}^3}{a_\text{A}} = \frac{p_\text{B}^3}{p_\text{A}} \cdot \frac{p_0}{p_0^3} = \frac{p_\text{B}^3}{p_\text{A}} \cdot (p_0)^{-2} \propto \frac{p_\text{B}^3}{p_\text{A}} \tag{10.53}$$

The expansion of ideal gas mixture at constant temperature means that all partial pressures are reduced by a factor of 5 ($pV = $ constant; Boyle's law):

$$Q = \frac{\left(\frac{p_\text{B}}{5}\right)^3}{\left(\frac{p_\text{A}}{5}\right)} \cdot (p_0)^{-2} = \left(\frac{p_\text{B}^3}{p_\text{A}}\right) \cdot (p_0)^{-2} \cdot \left(\frac{1}{5^2}\right) = \frac{K}{25} \tag{10.54}$$

The reaction quotient (Q) is only 4% of the equilibrium constant (K), which implies that the reaction will continue towards the product side (formation of great many B-molecules) until Q is equal to K.

23. The equilibrium constant (K) is given by:

$$K = \exp\left(\frac{-\Delta G_r^0}{RT}\right) \tag{10.55}$$

where ΔG_r^0 is:

$$\Delta G_r^0 = \sum_i v_i \Delta G_f^0 = \Delta G_f^0(HCl) - \frac{1}{2} \cdot \left(\Delta G_f^0(H_2) + \Delta G_f^0(Cl_2)\right) = \tag{10.56}$$
$$= -95300 + 0 + 0 = -95300 \text{ J mol}^{-1}$$

The equilibrium constant becomes $\exp(-95300/(8.314 \cdot 298)) = 5.2 \cdot 10^{16}$, which means that the reactants are is essentially completely transformed into product (HCl) molecules.

24. The equilibrium reaction is $CH_4(g) \leftrightarrows H_2(g) + C(s)$:

(i) The reaction standard Gibbs free energy is:

$$\Delta G_r^0 = \Delta H_r^0 - T\Delta S_r^0 \tag{10.57}$$

$$\Delta G_r^0 = -\Delta G_f^0(CH_4) = T\Delta S_f^0(CH_4) - \Delta H_f^0(CH_4) = \tag{10.58}$$
$$-298 \cdot 80.67 + 74850 = 50810 \text{ J mol}^{-1}$$

which is used to calculate K at 298 K:

$$K = \exp\left(\frac{-\Delta G_r^0}{RT}\right) = \exp\left(\frac{-50810}{8.314 \cdot 298}\right) = 1.24 \cdot 10^{-9} \tag{10.59}$$

(ii) The van't Hoff equation is used to calculate the equilibrium constant at 323 K:

$$\frac{d \ln K}{dT} = \frac{\Delta H_r^0}{RT^2} \tag{10.60}$$

$$\ln K(323 \text{ K}) = \ln K(298 \text{ K}) - \frac{\Delta H_r^0}{R}\left(\frac{1}{323} - \frac{1}{298}\right) = -18.17 \tag{10.61}$$
$$K(323 \text{ K}) = 1.28 \cdot 10^{-8}$$

(iii) The degree of conversion (α) is calculated as follows. First, the amount of unreacted CH_4 is $(1 - \alpha)$; the amounts of formed products are 2α (H_2) and α (C). Dalton's law can be used to calculate the partial pressures of H_2 and CH_4:

$$p_{H2} = \frac{2\alpha}{1+\alpha} \cdot P_{tot}; p_{CH4} = \frac{1-\alpha}{1+\alpha} \cdot P_{tot} \tag{10.62}$$

The equilibrium constant can be written:

$$K = \frac{a_{H2}^2 \cdot a_C}{a_{CH4}} = \frac{p_{H2}^2/(p^0)^2}{p_{CH4}/p^0} = \frac{p_{H2}^2}{p_{CH4} \cdot p^0} \tag{10.63}$$

which combined with Eq. (10.62) gives the degree of conversion of the methane molecules (α):

$$K = \frac{\left(\frac{2\alpha}{1+\alpha} \cdot P_{tot}\right)^2}{\left(\frac{1-\alpha}{1+\alpha} \cdot P_{tot} \cdot p^0\right)} = \frac{4\alpha^2 \cdot P_{tot}}{(1+\alpha) \cdot (1-\alpha) \cdot p^0} \Rightarrow$$

$$4\alpha^2 \cdot P_{tot} - Kp^0 (1 - \alpha^2) = 0 \tag{10.64}$$

$$\alpha = \sqrt{\frac{Kp^0}{4p_{tot} + Kp^0}} = \sqrt{\frac{1.24 \cdot 10^{-9} \cdot 1}{4 \cdot 0.01 + 1.24 \cdot 10^{-9} \cdot 1}} \ (\text{bar/bar}) = 1.76 \cdot 10^{-4}$$

25. The phase diagram in Fig. 10.3 shows the present phases and the eutectic point.

The lever rule provides information about the amount of each phase in the two-phase regions. Let us take a close look at point I. The molar fraction of

Fig. 10.3 Phase diagram with phases, eutectic point and three applications of the lever rule

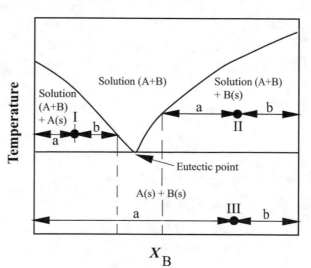

A(s) is $b/(a + b)$; the other phase, the solution of A + B with composition x_B obtained by the intersection of the vertical line through the x-axis, has an overall molar fraction of $a/(a + b)$. Point II: molar fraction of B(s) is $a/(a + b)$, and the molar fraction of the solution phase is $b/(a + b)$. Point III: the molar fraction of A(s) is $b/(a + b)$, and the molar fraction of B(s) is $a/(a + b)$.

Chapter 11
Mathematics Useful
for the Thermodynamics

A state function (e.g. the internal energy, U) can be a function of three *independent variables* x, y and z:

$$F = f(x, y, z) \tag{11.1}$$

The *total differential* of F is equal to the sum of the products of the three *partial derivatives* and the respective differentials of the independent parameters:

$$dF = \left(\frac{\partial F}{\partial x}\right)_{y,z} dx + \left(\frac{\partial F}{\partial y}\right)_{x,z} dy + \left(\frac{\partial F}{\partial z}\right)_{x,y} dz \tag{11.2}$$

The partial derivatives are similar to conventional derivatives; the only difference being that two independent variables are assumed to be constant, while F is differentiated with respect to the third independent variable. The procedure is further demonstrated by the following example:

$$F = 3x^2 + xy + z^2 y$$
$$\left(\frac{\partial F}{\partial x}\right)_{y,z} = 6x + y; \ \left(\frac{\partial F}{\partial y}\right)_{x,z} = x + z^2; \ \left(\frac{\partial F}{\partial z}\right)_{x,y} = 2zy \tag{11.3}$$

where the subscripts indicate the independent variables that are kept constant. The order of the derivation of the *second derivatives* is unimportant according to the *Schwarz theorem*, which, applied to $F(x,y,z)$, yields the following three equations:

© The Author(s), under exclusive license to Springer Nature Switzerland AG 2020
U. W. Gedde, *Essential Classical Thermodynamics*, SpringerBriefs in Physics,
https://doi.org/10.1007/978-3-030-38285-8_11

$$\left(\frac{\partial\left(\frac{\partial F}{\partial x}\right)_{y,z}}{\partial y}\right)_{x,z} = \left(\frac{\partial\left(\frac{\partial F}{\partial y}\right)_{x,z}}{\partial x}\right)_{y,z} \tag{11.4}$$

$$\left(\frac{\partial\left(\frac{\partial F}{\partial x}\right)_{y,z}}{\partial z}\right)_{x,y} = \left(\frac{\partial\left(\frac{\partial F}{\partial z}\right)_{x,y}}{\partial x}\right)_{y,z} \tag{11.5}$$

$$\left(\frac{\partial\left(\frac{\partial F}{\partial y}\right)_{x,z}}{\partial z}\right)_{x,y} = \left(\frac{\partial\left(\frac{\partial F}{\partial z}\right)_{x,y}}{\partial y}\right)_{x,z} \tag{11.6}$$

Applying Eqs. (11.4, 11.5, and 11.6) to the function shown in Eq. (11.3) demonstrates the Schwarz theorem:

$$\left(\frac{\partial\left(\frac{\partial F}{\partial x}\right)_{y,z}}{\partial y}\right)_{x,z} = 1 \wedge \left(\frac{\partial\left(\frac{\partial F}{\partial y}\right)_{x,z}}{\partial x}\right)_{y,z} = 1$$

$$\left(\frac{\partial\left(\frac{\partial F}{\partial x}\right)_{y,z}}{\partial z}\right)_{x,y} = 0 \wedge \left(\frac{\partial\left(\frac{\partial F}{\partial z}\right)_{x,y}}{\partial x}\right)_{y,z} = 0 \tag{11.7}$$

$$\left(\frac{\partial\left(\frac{\partial F}{\partial y}\right)_{x,z}}{\partial z}\right)_{x,y} = 2z \wedge \left(\frac{\partial\left(\frac{\partial F}{\partial z}\right)_{x,y}}{\partial y}\right)_{x,z} = 2z$$

In some cases, an equation is available relating the variables x, y and z:

$$x = f(y, z) \tag{11.8}$$

One example of this is the *ideal gas law*, which for a *closed system* (n is constant) is given by:

$$pV = nRT$$
$$p = nR \cdot \frac{T}{V} \Leftrightarrow p = f(T, V)$$
$$V = nR \cdot \frac{T}{p} \Leftrightarrow V = f(T, p) \tag{11.9}$$
$$T = \frac{1}{nR} \cdot pV \Leftrightarrow T = f(p, V)$$

Table 11.1 Functions and first derivatives

F (state function)	dF/dx
x^n	$n \cdot x^{n-1}$
a (a constant)	0
$\ln x$	$1/x$
$a x^n$	$a \cdot n \cdot x^{n-1}$
e^x	e^x

A state function such as U may be expressed either as a function of T and V, which is most common, or as a function of T and p or of V and p. Generally, this means that F is dependent only on *two independent variables* according to one of the following expressions:

$$F = f(x, y) \Leftrightarrow dF = \left(\frac{\partial F}{\partial x}\right)_y dx + \left(\frac{\partial F}{\partial y}\right)_x dy \qquad (11.10)$$

$$F = f(x, z) \Leftrightarrow dF = \left(\frac{\partial F}{\partial x}\right)_z dx + \left(\frac{\partial F}{\partial z}\right)_x dz \qquad (11.11)$$

$$F = f(y, z) \Leftrightarrow dF = \left(\frac{\partial F}{\partial y}\right)_z dy + \left(\frac{\partial F}{\partial z}\right)_y dz \qquad (11.12)$$

The choice of which of these equations should be used depends on what is readily measured. Because F is a *state function*, the actual path between the start (x_1, y_1, z_1) and the goal (x_2, y_2, z_2) is not decisive:

$$\Delta F = F(x_2, y_2, z_2) - F(x_1, y_1, z_1) \qquad (11.13)$$

However, as indicated in Chap. 1, the ΔF is calculated by assuming that there is a *reversible path* between the given states prescribed by (x_1, y_1, z_1) and (x_2, y_2, z_2). As long as the actual parameter values of the start and the end are known, ΔF can be derived.

Table 11.1 presents a list of derivatives which are commonly present in thermodynamics equations. By combining these with some very useful methods, most of the problems concerned with the derivatives can be solved. Starting with the derivative of a composite function which is the product of the two functions f and g:

$$F = f(x, y) \cdot g(x, y)$$
$$\left(\frac{\partial F}{\partial x}\right)_y = \left(\frac{\partial f}{\partial x}\right)_y \cdot g + \left(\frac{\partial g}{\partial x}\right)_y \cdot f \qquad (11.14)$$

An analogous equation is obtained for the derivative with respect to the other independent variable, y. The same can be claimed about Eqs. (11.15, 11.16, 11.17, and 11.18). The derivative of a composite function where one function (f) is divided by another function (g) is given by:

$$F = f(x,y)/g(x,y)$$

$$\left(\frac{\partial F}{\partial x}\right)_y = \frac{\left(\frac{\partial f}{\partial x}\right)_y \cdot g - \left(\frac{\partial g}{\partial x}\right)_y \cdot f}{g^2} \tag{11.15}$$

The derivative of an exponential of the function is given by:

$$F = (f(x,y))^n$$

$$\left(\frac{\partial F}{\partial x}\right)_y = n \cdot f^{n-1} \cdot \left(\frac{\partial f}{\partial x}\right)_y \tag{11.16}$$

The famous *chain rule*, which was first proposed by Gottfried Wilhelm Leibniz (German mathematician) in the seventeenth century, is:

$$F = f(u(x,y))$$

$$\left(\frac{\partial F}{\partial x}\right)_y = \left(\frac{\partial F}{\partial u}\right) \cdot \left(\frac{\partial u}{\partial x}\right)_y \tag{11.17}$$

And the chain rule can be used, for example, in the following way:

$$F = e^{x^2+y}$$

$$F = e^u; u = x^2 + y$$

$$\left(\frac{\partial F}{\partial x}\right)_y = e^u \cdot 2x = 2x \cdot e^{x^2+y} \tag{11.18}$$

The chain rule is probably the most useful method to reveal derivatives on the basis of a few simple principles. Two other useful expressions are:

$$\frac{dy}{dx} = \frac{1}{\left(\frac{dx}{dy}\right)} \tag{11.19}$$

$$f(x,y,z) = 0$$

$$\left(\frac{\partial x}{\partial y}\right)_z \cdot \left(\frac{\partial y}{\partial z}\right)_x \cdot \left(\frac{\partial z}{\partial x}\right)_y = -1 \tag{11.20}$$

The latter is referred to as the *Euler chain rule*.

The solution of an expression like Eq. (11.2) takes the form $F = (x,y,z)$, and this requires *integration*, i.e. a procedure which is the reverse of differentiation. A simple example is:

Table 11.2 Functions and their integral functions

$F(x)$	$\int F(x)dx$
x^n	$1/(n+1)\cdot x^{n+1} + C$
$1/x$	$\ln x + C$
e^x	$e^x + C$
e^{-x}	$-e^{-x} + C$
$\ln x$	$x \cdot (\ln x - 1) + C$

$$\frac{dy}{dx} = x^2 \Rightarrow dy = x^2 dx \Rightarrow \int dy = \int x^2 dx$$

$$y = \frac{x^3}{3} + C \Leftrightarrow \frac{dy}{dx} = \frac{3x^3}{3} = x^2$$

(11.21)

The solution to the differential equation contains C, a constant which cannot be given a precise value unless a pair of x, y values are known. The full solution, assuming that this point is (x_1, y_1), is given by:

$$y_1 = \frac{x_1^3}{3} + C \Rightarrow C = y_1 - \frac{x_1^3}{3}$$

$$y = \frac{x^3}{3} + y_1 - \frac{x_1^3}{3}$$

(11.22)

Table 11.2 presents a set of solutions of some integrals relevant for thermodynamics. Polynomials, exponentials and logarithms are common in classical thermodynamics equations.

The integration of more complex expressions requires the use of some additional methods. These methods together with the solutions shown in Table 11.2 make it possible to solve most of the differential equations used in classical thermodynamics. These methods are here presented in a brief form, in several cases by a demonstrative example.

A *constant factor* is readily dealt with according to:

$$y = \int ax^2 dx = a \int x^2 dx = a \cdot \frac{x^3}{3} + C$$

(11.23)

Integration of *sums* and *differences* are treated as in the following example:

$$y = \int (3 + x^2 + x^4)dx = \int 3dx + \int x^2 dx + \int x^4 dx$$

$$y = 3x + \frac{x^3}{3} + \frac{x^5}{5} + C$$

(11.24)

More complex expressions can be dealt with by *substituting* an expression with a new variable (denoted z in this case) according to the following example:

$$
\begin{aligned}
y &= \int \sqrt{2x-1}\,dx \\
z &= 2x - 1 \wedge dz = 2dx \Rightarrow dx = dz/2 \\
y &= \int \sqrt{z} \cdot \frac{dz}{2} = \frac{1}{2} \cdot \frac{z^{\frac{3}{2}}}{\frac{3}{2}} + C = \frac{z^{\frac{3}{2}}}{3} + C = \frac{(2x-1)^{\frac{3}{2}}}{3} + C
\end{aligned}
\tag{11.25}
$$

It is important to establish a relationship between the differential of the new parameter (dz) and dx. Another occasionally useful technique is to use the method referred to as *integration by parts*:

$$
\begin{aligned}
\int u\,dv &= uv - \int v\,du \\
y &= \int xe^x dx = \int x(e^x dx) \\
u &= x \Rightarrow du = dx \wedge dv = e^x dx \Rightarrow v = e^x \\
y &= \int x(e^x dx) = xe^x - \int e^x dx \Rightarrow y = e^x(x-1) + C
\end{aligned}
\tag{11.26}
$$

This method is useful if the integral in the right-hand part of the equation is readily soluble by a standard method. One of the most powerful methods is the method whereby the integral is expressed simple factors, which is referred to as *integration by partial fractions*. The following integral is solved, for example, in this way:

$$
y = \int \frac{dx}{x^2 + 2x - 3}
\tag{11.27}
$$

The denominator is first expressed in factors according to:

$$
\begin{aligned}
x^2 + 2x - 3 &= 0 \Rightarrow (x-1)(x+3) = 0 \\
y &= \int \frac{dx}{(x-1)(x+3)} = \int \left(\frac{A}{x-1} + \frac{B}{x+3} \right) dx \\
A(x+3) &+ B(x-1) = 1 \\
x\text{-terms} &: A = -B \\
\text{numbers} &: 3A - B = 1 \Rightarrow A = 1/4; B = -1/4
\end{aligned}
\tag{11.28}
$$

which provides the input to solve the integral according to:

$$y = \frac{1}{4} \cdot \left(\int \frac{dx}{x-1} - \int \frac{dx}{x+3} \right) = \frac{1}{4} \cdot \left(\ln(x-1) - \ln(x+3) \right) + C$$

$$y = \frac{1}{4} \cdot \ln \left(\frac{x-1}{x+3} \right) + C$$

(11.29)

So far all the treatment of integral calculus has been concerned with what are called *general integrals*. These are functions, and typically they all have a term C, indicating the undetermined constant. If the starting conditions are known, the constant C can be determined, but this does not alter the fact that the general integral remains a function. *Definite integrals* are integrals with upper and lower limits, and they are not functions; they are represented by scalar values. A simple example of the two integral types is shown in:

$$(a)\ y = \int x^2 dx = \frac{x^3}{3} + C$$

$$(b)\ y = \int_1^3 x^2 dx = \left[\frac{x^3}{3} \right]_1^3 = \frac{27}{3} - \frac{1}{3} = \frac{26}{3}$$

(11.30)

The expression given in Eq. (11.30a), is a function (general integral), whereas the equation in Eq. (11.30b), with the lower and upper limits 1 and 3, respectively, and the value 26/3 is a definite integral.

A few other mathematical tricks are used in classical thermodynamics, but we leave that to you to find out. Recommended reading, covering essentially all the mathematics needed for classical thermodynamics, is the fantastic text by Thompson and Gardner (1998). This very affordable book is based on Silvanus P. Thompson book first published in 1910 under pseudonym "F. R. S.", Fellow of the Royal Society. The identity of the author was not revealed until after the death of Thompson. The title of the book given by Thompson says something, "Calculus made easy – being a very-simplest introduction to those beautiful methods of reckoning which are generally called by the terrifying names of the differential calculus and the integral calculus", a great text which has been modernized in a gentle fashion by Martin Gardner.

References

Friedrich, H. E., & Mordike, B. L. (2006). *Magnesium technology metallurgy, design, data, applications* (pp. 80–124). Berlin: Springer.

Gedde, U. W., & Hedenqvist, M. S. (2019). *Fundamental polymer science.* Berlin/New York/London: Springer Nature.

Hall, D. L., Sterner, S. M., & Bodner, R. J. (1988). Freezing point depression of NaCl-KCl-H_2O solutions. *Economic Geology, 83,* 197.

Schmidt, A. (2019). *Technical thermodynamics for engineers.* Berlin/New York/London: Springer Nature.

Tellefsen, T. A., Larsson, A., Løvvik, O. M., & Aasmundsveit, K. (2012). Au-Sn SLID bonding—properties and possibilities. *Metallurgical and Materials Transactions B, 43,* 397.

Thompson, S. P., & Gardner, M. (1998). *Calculus made easy.* London: Macmillan.

© The Author(s), under exclusive license to Springer Nature Switzerland AG 2020   101
U. W. Gedde, *Essential Classical Thermodynamics*, SpringerBriefs in Physics,
https://doi.org/10.1007/978-3-030-38285-8

Index

Printed in the United States
By Bookmasters